Heron Streaming

Huijun Wu • Maosong Fu

Heron Streaming

Fundamentals, Applications, Operations,
and Insights

 Springer

Huijun Wu
Twitter (United States)
San Francisco, CA, USA

Maosong Fu
Twitter (United States)
San Francisco, CA, USA

ISBN 978-3-030-60093-8   ISBN 978-3-030-60094-5   (eBook)
https://doi.org/10.1007/978-3-030-60094-5

This Springer imprint is published by the registered company Springer Nature Switzerland AG.
The registered company address is: Gewerbestrasse 11, 6330 Cham, Switzerland

To the Heron community.

Foreword

In 2017 when I was on vacation, Maosong reached out to me and asked if I would be interested in joining the Real-Time Compute (RTC) team as the tech lead. I did not give him an answer at the time. However, soon some other things happened (long story and there are always plenty of stories at Twitter), and I finally decided to join the team and the project. "Life is like a box of chocolates. You never know what you're gonna get," said Forrest Gump in one of my favorite movies. I guess "when" might be another thing you never know. Anyways, that was the time I started working with Huijun, Maosong, and the great team.

Stream processing was quite new for me and very different from what I had been working on earlier (such as games, video players, and web and backend services). In the RTC team, I learned a lot about the technology from my teammates Huijun, Neng, Mark, and Maosong, as well as ex-teammates Sanjeev and Karthik, and the whole Apache Heron (Incubating) community. Many of them were Apache Storm developers before and founding members of the Apache Heron project, so I got a lot of insider or original information for my study. I feel I was really lucky to join at a great time. Without their help, it would have been a much bigger challenge for me.

It should not be hard to imagine that stream processing is widely used at Twitter. After Twitter acquired BackType (the original developer of Apache Storm), a lot of streaming jobs have been built for Twitter's core business and other data analytics tasks on top of it. Later, the RTC team decided to start a new re-architecture project, Heron, to address some serious concerns and pain points learned from operating Storm at Twitter. Among the many changes, the new project was aimed at being single-tenant, highly modularized, and customizable. The project was introduced to the public at the SIGMOD 2015 and open-sourced in 2016. In 2018, Twitter donated Heron to Apache. The Heron project is the outcome of a lot of previous experience and hard work from many people, and this book is a good summary of what the authors know (and they are definitely qualified for the task).

After leaving Twitter around the end of 2019 (another long story not for today) to join a data analytics start-up, Amplitude, I spent some time to step back and asked myself two questions from an outsider point of view: As Heron is designed to address the issues with Twitter's use cases, how useful is it for other people or companies? What is missing for Heron to be more popular? After some reflection, my thoughts are as follows:

- Yes, I do believe Heron could be useful for other people. Twitter is unique, but not that unique. It is likely that other people will experience similar pains, but everyone would handle them very differently. Furthermore, stream processing use cases can be complicated, and there are no one-size-fits-all solutions. Therefore, having different options to choose from can be valuable for many systems. This could be extremely powerful because with Heron you can replace modules (e.g., scheduler and packing algorithms) with your own implementation when necessary.
- There is a consequence of the highly modularized and configurable architecture: the learning curve to adopt could be steep. After all, not many people are as lucky as me to have access to a lot of "undocumented" information.

Now, with this new book by my ex-teammates Huijun and Maosong, you could be as lucky as me! They share a lot of their knowledge about Heron in this book, from the original design goal, to how to build topologies (Heron and Storm's term for streaming jobs), how to operate, and how to extend Heron to suit your use cases better. In case you are looking for options to improve your systems or are simply curious about how Heron works or its differences from Apache Storm, this book could be right for you. I really wish that the book were available when I started to work on the project. My ex-teammates would have saved a lot of time spent answering my endless questions (many thanks!).

In addition to current and future Heron users, I believe this book (and the Heron project) could also be useful for people who are building (or are interested in building) their own streaming engines. Since Heron is highly modularized, it could be easier for you to study how a complex streaming engine is built piece by piece, and the book could be a good guide for you to navigate through the project.

Ex-Technical Lead of Twitter's   Ning Wang
Real-Time Compute Team
Coauthor *Grokking Streaming Systems*
Silicon Valley, CA, USA
August 2020

Preface

Stream processing is a must-have capability of the modern data platform for leading Internet services such as Twitter, Facebook, and YouTube. Many stream processing systems have been developed to fulfill the real-time data processing requirement, among which Apache Heron (Incubating) is a new and promising project.

The Heron project originated at Twitter, Inc., became open source in the summer of 2016, and joined Apache Incubator in the spring of 2018. Heron is entirely new, and many users, operators, and developers need a book to start with, which has inspired the authors to write this book.

Who This Book Is For

This book targets readers who want to process streaming data based on the Heron project and who fall into the following two categories:

- Software Engineer (SWE): Heron provides two sets of programming interfaces: the application programming interface (API) for SWEs to develop topologies and the service provider interface (SPI) for SWEs to extend Heron capabilities. Therefore, the SWEs further fall into the following two subcategories:

  - SWE for topology (developers): This book will describe how to migrate an existing Apache Storm topology and how to write topologies with Heron APIs (including both the low-level Topology API and the high-level Streamlet API in Java, Python, and Scala).
  - SWE for Heron (contributors): This book will describe the Heron architecture, the components, and the interactions between them, which paves the way for future Heron contributors.

- Site Reliability Engineer (SRE) (operators): The Heron topologies have their life cycles that need SREs to manipulate and monitor. This book will describe how to manipulate the topology life cycle and how to monitor the topology running status.

How This Book Is Organized

This book is organized into four parts:

- Part I provides basic knowledge about stream processing, Apache Storm, and Apache Heron (Incubating). After Heron joined the Apache family, source code became mandatory in its release. This part will also introduce the Heron source repository and is suggested for all readers.

  - Chapter 1 introduces the stream processing basics and the history behind Apache Storm and Apache Heron (Incubating).
  - Chapter 2 introduces the Heron basics, including concepts, architecture, runtime behaviors, etc., and demonstrates the first topology in local mode.
  - Chapter 3 introduces the Heron code repository and explores how code is wired together.

- Part II describes two data models to write Heron topologies and often-used topology features, including stateful processing. This part is suggested for SWEs (developers) who write topology with Heron APIs.

  - Chapter 4 presents procedures to migrate an existing Apache Storm topology to a Heron topology.
  - Chapter 5 introduces the Heron Topology API to write topologies and studies how the tuples and streams are transmitted in the topology.
  - Chapter 6 discusses topology features, including delivery semantics, windowing, etc.
  - Chapter 7 presents the Heron Streamlet API, which is a higher level than the Topology API and is inspired by functional programming.

- Part III describes Heron tools, including the command-line interface, user interface, etc., to manage a single topology or multiple topologies in a data center. This part is suggested for SREs (operators) who deploy and manage running jobs.

  - Chapter 8 introduces the topology life cycle and how to use the Heron CLI to manipulate a topology.
  - Chapter 9 introduces Heron tools for operators to manage multiple topologies in a data center.

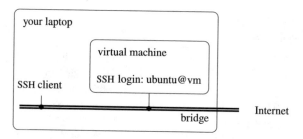

Fig. 1 Experiment environment

- Part IV describes the Heron source code and how to customize or extend Heron. This part is suggested for SWEs (contributors) who would like to contribute code to the Heron repository and who are curious about Heron insights.

  - Chapter 10 introduces Heron modules and explores how these modules work together.
  - Chapter 11 discusses the Heron metrics' flow and how to route metrics to any destination through a metrics sink.
  - Chapter 12 discusses Heron Scheduler and how to write a scheduler to run Heron jobs on a new resource pool.
  - Chapter 13 introduces the newly developed features and desired features from the community.

This book includes multiple demonstration projects, which are summarized in Table 1. The project codes are available at https://github.com/StreamProcessingWithHeron. The Index at the back of this book lists concept and terminology definitions.

What You Need for This Book

For this book, you will need a Ubuntu box. You may run a Ubuntu virtual machine if you do not have a Ubuntu machine, as shown in Fig. 1. All the examples in this book have been verified on Ubuntu Server 20.04[1] with the latest Apache Heron (Incubating) release candidate 0.20.3-incubating-rc7. The examples may also run on a Macintosh with minor adjustments.

You will need to know Java to understand most of the example code and Bash commands for Linux operations in this book. You may also need to have a basic knowledge of Python and Scala for some of the examples.

---

[1] Authors use https://releases.ubuntu.com/20.04/ubuntu-20.04-live-server-amd64.iso to establish the environment.

Table 1 Demonstration topologies

Chapter	Project name	Topology name
Chapter 2	heron	hello-world-topology
Chapter 4	heron-starter	storm-migration-java
Chapter 4	heron-eco	yaml-topology
Chapter 5	heron-topology-java	my-java-topology
Chapter 5	heron-topology-py	my-python-topology
Chapter 6	heron-topology-java-at-least-once	my-java-topology-at-least-once
Chapter 6	heron-topology-java-effectively-once	my-java-topology-effectively-once
Chapter 6	heron-topology-java-window	my-java-topology-window
Chapter 7	heron-streamlet-java	my-java-streamlet
Chapter 7	heron-streamlet-py	my-python-streamlet
Chapter 7	heron-streamlet-scala	my-scala-streamlet
Chapter 8		my-test-topology-1, my-test-topology-2
Chapter 9		my-test-topology-3, my-test-topology-4
Chapter 10	heron-proto	
Chapter 11	heron-metrics-sink	
Chapter 12	heron-scheduler	

Typographical Conventions

We use the following typographical conventions in this book:

Italic
> Indicates names, rather than codes or commands, including new terms, URLs, email addresses, project names, component names, stream names, instance names, and filenames.

`Constant width`
> Shows code listings, as well as in-line codes, including variables, functions, classes, objects, keywords, and statements.

`Constant width bold`
> Shows executable binaries and commands.

`Constant width italic`
> Indicates text that ought to be supplanted with values determined by contexts, such as command and code parameters, arguments, and optional values depending on application scenarios.

TIP

This box provides a tip or suggestion.

> **NOTE**
>
> This box indicates a general note.

Acknowledgments

Twitter, Inc. is where Heron was born. The authors would like to thank Twitter for its continuous investment in the Heron project. Twitter is one of the leading companies in the stream processing area. Twitter donated Apache Storm, the predecessor of Heron. Without the wisdom and generosity of Twitter, there would be neither the de facto stream processing engine standard, Apache Storm, nor the next generation system, Heron. The authors would like to thank current and former colleagues Yao Li, Xiaoyao Qian, Dmitry Rusakov, Neng Lu, Ning Wang, Yaliang Wang, Runhang Li, Bill Graham, Cong Wang, and Karthik Ramasamy for the close collaboration in the Real-Time Compute team. They would especially like to thank Karthik Ramasamy, who has been leading the Heron project for years. The authors would like to thank Sree Vaddi, Josh Fischer, Sanjeev Kulkarni, Ashvin Agarwal, Avrilla Floratou, Boyang Jerry Peng, Thomas Cooper, and Faria Kalim for their continuous contribution to the Heron project. Finally, the authors would like to thank Ralf Gerstner from Springer for guidance through the publishing process.

Silicon Valley, CA, USA   Huijun Wu
Silicon Valley, CA, USA   Maosong Fu
August 2020

Contents

About the Authors

Huijun Wu is an engineer at Twitter, Inc. He has been working on the Heron project since the summer of 2016 when Twitter open-sourced the Heron code and is a founding member of Apache Heron (Incubating). He has also worked for Microsoft, ARRIS, and Alcatel Lucent. He is also the coauthor with Dijiang Huang of the book *Mobile Cloud Computing: Foundations and Service Models* (Morgan Kaufmann). He received a Ph.D. from the School of Computing Informatics and Decision Systems Engineering at Arizona State University.

Maosong Fu is the engineering manager for the Real-Time Compute team at Twitter. Previously, he was the technical lead for Heron at Twitter and has worked on various components of Heron, including Heron Instance, Metrics Manager, Heron Scheduler, etc. He is the author of publications on the topic of distributed areas. He received a master's degree from Carnegie Mellon University and a bachelor's degree from Huazhong University of Science and Technology.

Part I
Heron Fundamentals

Chapter 1
Stream Processing

With the rapid development of the Internet, especially with the wide application of social networks, the Internet of things (IoT), cloud computing, and a variety of sensors in recent years, unstructured data characterized by a large number, many varieties, and strong timeliness has emerged. The importance of data is becoming more prominent. Traditional data storage and analysis techniques find it difficult to process large amounts of unstructured information in real time, thus giving rise to the concept of "Big Data."

The wide application of big data technology makes it a critical supporting technology that leads to the technological advancement of many industries and promotes revenue growth. According to the timeliness of data processing, big data processing systems can be divided into two types: batch processing and stream processing.

Batch processing

Batch processing is ideal for calculations that require access to a full set of records. For example, when calculating totals and averages, you must treat the data set as a whole. Batch processing is a good candidate for processing large amounts of historical data since it takes much time, but it is not good for real-time applications requiring low latency. Apache Hadoop is a typical processing framework for batch processing. Hadoop includes the Hadoop distributed file system (HDFS) and the batch processing engine MapReduce.

Stream processing

The stream processing system calculates the data that enters the system at any time. This is an entirely different approach than the batch mode. The stream processing method does not need to perform operations on the entire data set but performs operations on each data item transmitted through the system. Running MapReduce is not a good idea for stream processing due to the I/O bottlenecks of HDFS (or any other file system for that matter). HDFS is suited for long reads and short writes like applications. For streaming applications, it is better to process the data in memory (i.e., RAM) rather than put it on a disk, etc.

Storm, Spark Streaming, and Flink are good examples of stream processing applied to big data. If you take Storm as an example, it can orchestrate the directed acyclic graph (DAG) named topology in the framework. These topologies describe the different transformations or steps that need to be performed on each incoming data tuple as it enters the system [2].

Reducing processing delays between data producers and consumers has been the main driver of the evolution of modern computing architectures. As a result, computing frameworks for real-time and low-latency processing, such as Lambda and Kappa, have been born, and such hybrid architectures complement each other, connecting traditional batch processing and stream processing. We will explore the two crucial data processing architectures Lambda and Kappa in Sect. 1.1. We will then focus on stream processing in Sect. 1.2 and introduce the Heron project background in Sect. 1.3. In Sect. 1.4, we will summarize the state-of-the-art stream processing tools.

1.1 Big Data Processing

In the application scenarios of Internet/mobile Internet and IoT, complex business requirements such as personalized service, improved user experience, intelligent analysis, and decision-making in the field put forward higher expectations for big data processing technology. To meet these needs, big data processing systems must return processing results in milliseconds or even microseconds. Such high concurrency and high real-time application requirements pose serious challenges for big data processing systems. Two architectures have emerged to address these challenges, as discussed in the following text.

1.1.1 Lambda Architecture

When using a huge data set, queries can take a long time. They cannot be executed in real time and typically require parallel processing on the whole data set. The results are then stored separately in a new data set for querying.

One disadvantage of this method is that it causes a delay—if the processing takes hours, the result returned by the query may be the result of data that was available several hours ago. It is best to get some real-time results (perhaps less accurate) and then combine those results with the batch analysis results.

Lambda data architecture is a must-have architecture for every company's big data platform. It solves the need for a company's big data batch offline processing and real-time data processing. The core of a typical Lambda architecture is divided

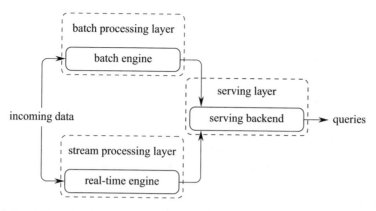

Fig. 1.1 Lambda architecture

into three layers [1]: the batch processing layer, the stream processing layer, and the serving layer, as shown in Fig. 1.1.

1.1.1.1 Batch Processing Layer

New data is continually being used as a source of data systems, which is fed into both the batch and stream processing layers. The batch processing layer stores the data set, pre-calculates the query function on the data set, and builds a view corresponding to the query. The batch processing layer can handle offline data very well, but there are many cases where data is generated in real time, and real-time query processing is required. For this situation, the stream processing layer is more suitable. Hadoop is the first big data processing framework, and for a long time, it has been used as a synonym for big data technology.

1.1.1.2 Stream Processing Layer

The batch processing layer handles the entire data set, while the stream processing layer handles the most recent incremental data stream. The stream processing layer only processes real-time data of small computational load, which leads to low latency. It emits output when watermarks are triggered, facilitating the serving layer to respond to queries quickly.

Apache Storm and the related Apache Kafka are typically used in this layer. Apache Storm makes it easy to reliably process unbounded streams of data, doing for real-time processing what Hadoop does for batch processing [9]. Kafka is a fast, scalable, high-throughput, fault-tolerant distributed publish and subscribe messaging system, which is often used as a data source and sink for Storm jobs.

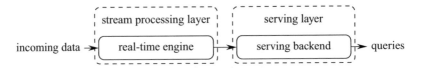

Fig. 1.2 Kappa architecture

1.1.1.3 Serving Layer

The serving layer is used to merge the result data sets in the batch view and real-time view into the final data set. The query presents this final result in a combination of batch view and real-time view.

Druid is one of the example technologies used in the serving layer, which is a data store designed for high-performance slice-and-dice analytics on large data sets [3]. Druid is most often used as a backend for highly concurrent application programming interfaces (APIs) that need fast aggregations.

1.1.2 Kappa Architecture

One obvious problem with the Lambda architecture is the need to maintain two sets of code that run on both batch and real-time processing systems and yield the same results. So the questions for the people who design such systems are: Why can we not improve the stream processing system so that it can handle these problems? Why can the streaming system not solve the problem of full data processing? The natural distributed characteristics of stream processing are destined to be more scalable. Can you increase the amount of concurrency to handle massive historical data? Based on these considerations, Kappa was proposed. The Kappa and Lambda architectures share the same primary goals, but the Kappa architecture has one crucial difference: all data flows through one path, using a stream processing system.

The Kappa architecture is shown in Fig. 1.2. It focuses only on stream computing. For this architecture, data is collected in streams, and the stream processing layer puts the results into the serving layer for queries.

The idea is to integrate batch processing and stream processing in a single stream processing engine. This requires the collected data stream to be replayed from a new location or all at once. If the calculation logic is changed, a new flow calculation will be performed, the previous data will be quickly played back and recalculated, and the new results will be saved to the serving layer.

This architecture attempts to maintain only one piece of code, simplifying the code in the Lambda architecture that maintains both batch and stream processing layers. Also, the results of the query only need to be in a single serving layer, and there is no need to integrate batch and real-time results.

1.2 Big Data Stream Processing

Stream processing adheres to the basic notion that data values fall rapidly over time, such as real-time recommendations, anomalies or fraud detection, and real-time reporting. In the real-time recommendation scenario, when the user clicks, the system perceives the user's current interest, finds the content that is of interest to the user from the content library, and pushes it to the user, so that the user can discover what he/she is interested in in real time and thereby extend the user's active time. Conversely, when a user accesses an offline recommendation system, even with an algorithm having a higher accuracy rate, the user's interest may change over time. In real-time anomaly and fraud detection scenarios, taking an e-commerce system as an example, the system needs to monitor the user's behavior in real time to determine whether the user is a connoisseur. If the system is a batch processing system, the system transaction may have been completed when a problem is found.

Each tuple of the streaming data is processed when the tuple reaches the system, while the batch processing system loads the whole data set and processes it at intervals. A new operation called "window" was introduced to stream processing recently, which allows the stream processing system to accumulate the incoming data within a particular time interval marked by "watermark" and process the data in the window at one time. Stream processing has the following characteristics:

Quick Response

The rise of big data established that insights derived from processing data are valuable. Such insights are not all created equal. In some scenarios, the value of insights diminishes quickly with time. Stream processing was developed for such scenarios. The advantage of stream processing is that it provides insights within milliseconds to seconds. Applications and analytics react to events instantly. The lag time between "event happens" $\rightarrow$ "insight derived" $\rightarrow$ "action is taken" is significantly reduced. Actions and analytics are immediate, reflecting the data when it is still fresh, meaningful, and valuable.

Endless Event Sequence

Streaming data is a series of sequential, large, fast, and contiguous data. In general, a data stream can be thought of as a dynamic data set that grows indefinitely over time. The data set in stream processing is "endless," which has several significant effects. First, a complete data set can only represent the total amount of data that has entered the system so far. Second, a working data set may be more relevant and can only represent a single data item at a particular time. Third, the processing is event based, and there is no "end" unless explicitly stopped. Processing results are immediately available and continue to be updated as new data arrives.

Batch processing is suited for use cases where the response time is not a critical requirement. Typical use cases include offline ranking on the historical data and building an index on the historical data. However, stream processing is typically used for applications that require real-time interaction and real-time response. Typical application scenarios for stream processing include real-time recommendations, industrial IoT, and fraud detection.

Real-time recommendations

The system obtains the user's portrait, performs a match in the content database, finds the content that is similar to the user's portrait, and pushes it to the user. In order to solve the real-time nature, the user portrait is divided into two parts: a long-term interest label and a short-term interest label. The long-term interest label refers to the user's portrait that does not change very much or has a relatively low frequency of change, such as gender, age, geographical location, and consumption habits of the user. Since the long-term interest label does not require real time, it can be calculated offline. For short-term interest labels, the system records the user's behavior and topics of interest through the web or app and sends it to the message queue. For example, if a user continually refreshes "car"-related web pages, the system subscribes to the message queue and calculates the user's short-term interest labels in real time and then sorts them out, and the content-matched search returns the result to the user.

Industrial IoT

The user has a production line and wants to obtain real-time indicators on the production line or monitor the yield rate on the production line to prevent further losses. The user deploys the sensor on the lathe to collect the characteristics of the product in real time, which sends the original information of the feature to the message queue server. The stream processing system subscribes to the message queue to obtain the data of the product and calculates the characteristics of the product. The system calculates the yield rate by detection models and finally presents the result to the user in a personalized manner.

Fraud detection

It is assumed that the e-commerce system monitors the user's behavior in real time. Its data flow is similar to the industrial IoT: the user's behavior is sent to the message queue server, following which the system characterizes the user behavior in real time, and then the risk model is invoked to make judgments on the user's behavior. The risk model can detect if there is a problem with the user's behavior in real time and then issue an alarm to avoid more significant property damage.

1.3   From Apache Storm to Apache Heron (Incubating)

We saw in the previous sections that both the Lambda and Kappa architectures include a speed or stream processing layer, and the standard technology used is Apache Storm.

Apache Storm is a free and open-source distributed real-time computation system [9] written mainly in Java. BackType created Storm, which was acquired by Twitter, and then the project was open sourced. Storm uses "spouts" and "bolts" to define a processing graph or "topology" to process streaming data in real time. The initial release was on September 17, 2011. Storm became an Apache Top-Level Project in September 2014 and was previously in incubation since September 2013.

A Storm data model is based on a directed acyclic graph called "topology." The vertices of the graph are spouts and bolts. The edges of the graph are streams transmitting data between nodes. The topology acts as a data processing pipeline.

1.3.1   Motivation for Heron

Storm was the primary real-time processing platform at Twitter for several years. While we were operating Storm, we found that it had some limitations, which motivated us to develop Heron [7]:

Worker limitation

A Storm worker has a fairly complex design. The operating system schedules several instances of worker processes in a host. Inside the Java virtual machine (JVM) process, each executor is mapped to two threads scheduled using a pre-emptive and priority-based scheduling algorithm by the JVM. Since each thread has to run several tasks, the executor implements another scheduling algorithm to invoke the appropriate task, based on the incoming data. Such multiple levels of scheduling and their complex interaction often lead to uncertainty about when the tasks are being scheduled.

Furthermore, each worker can run different tasks. It is difficult to reason about the behavior and the performance of a particular task since it is not possible to isolate its resource usage. For resource allocation purposes, Storm assumes that every worker is homogeneous. This architectural assumption results in an inefficient utilization of allocated resources and often results in overprovisioning.

Nimbus issue

The Storm Nimbus (master node of cluster) performs several functions, including scheduling, monitoring, and distributing Java archives (JARs). It also serves as the metrics-reporting component for the system and manages counters for several topologies. Thus, the Nimbus component is functionally overloaded and often becomes an operational bottleneck for a variety of reasons, as follows.

First, the Nimbus scheduler does not support resource reservation and isolation at a granular level for Storm workers. Consequently, Storm workers that belong to different topologies running on the same machine could interfere with each other. Second, Storm uses ZooKeeper extensively to manage heartbeats from workers and supervisors. Such use of ZooKeeper limits the number of workers per topology, and the total number of topologies in a cluster, as at huge numbers, ZooKeeper becomes a bottleneck. Finally, the Nimbus component is a single point of failure.

Lack of backpressure

> Storm has no backpressure mechanism. If the receiver component is unable to handle incoming data/tuples, then the sender simply drops tuples. Backpressure refers to the buildup of data at a socket when buffers are full and are not able to receive additional data. No additional data packets are transferred until the bottleneck has been eliminated or the buffer has been emptied.

Efficiency

> In production, there were several instances of unpredictable performance during topology execution, which then led to tuple failures, tuple replays, and execution lag (the rate of data arrival exceeded the rate of processing by the topology). The most common causes for these reduced performance scenarios were suboptimal replays, long garbage collection cycles, and queue contention. To mitigate the risks associated with these issues, overprovisioning the allocated resources was often adopted. Such overprovisioning has obvious negative implications on infrastructure costs. See Sect. 13.1 for more details on topology tuning.

1.3.2   Heron Design Goals

Heron was built to fulfill an initial set of core design goals, including [6]:

Isolation

> Topologies should be process based rather than thread based, and each process should run in isolation for the sake of easy debugging, profiling, and troubleshooting.

Resource constraints

> Topologies should use only those resources that they are initially allocated and never exceed those limits. This makes Heron safe to run in a shared infrastructure.

Compatibility

> Heron is fully API and data model compatible with Storm, making it easy for developers to transit between systems. We will discuss how to migrate from Storm to Heron in Chap. 4.

Backpressure

> In a distributed system like Heron, there are no guarantees that all system components will execute at the same speed. Heron has built-in backpressure mechanisms to ensure that topologies can self-adjust in case components lag.

Performance

> Many of Heron's design choices have enabled it to achieve higher throughput and lower latency than Storm while also offering enhanced configurability to fine-tune potential latency/throughput trade-offs.

Semantic guarantees

> Heron provides support for at-most-once, at-least-once, and effectively-once processing semantics. We will see the details of these three semantic categories in Sect. 6.1.

Table 1.1 Apache mailing lists

Scope	Name
Development-related discussions	dev@heron.incubator.apache.org
Source code merge notifications	commits@heron.incubator.apache.org
User-related discussions	users@heron.incubator.apache.org
Report issues	issues@heron.incubator.apache.org

Efficiency

Heron was built to achieve all of the above with the minimum possible resource usage.

1.3.3 Join the Apache Heron (Incubating) Community

Heron was first announced on the Twitter Engineering Blog (TEB), "Flying Faster with Twitter Heron," in June 2015 [10], along with a paper, "Twitter Heron: Stream Processing at Scale," at ACM SIGMOD 2015 [7]. In May 2016, Heron was announced as open source in the TEB "Open Sourcing Twitter Heron" [12]. In February 2018, Twitter donated Heron to the Apache Software Foundation (ASF) as announced in the TEB "Leaving the Nest: Heron Donated to Apache Software Foundation" [11].

The current project website is https://heron.incubator.apache.org/, and the documentation is available at https://heron.incubator.apache.org/docs/getting-started-local-single-node. The project landing page in Apache is http://incubator.apache.org/projects/heron.html. There are four mailing lists for different purposes, as shown in Table 1.1.

For source code hosting, the community decided to continue with GitHub at https://github.com/apache/incubator-heron, as well as bug tracking at https://github.com/apache/incubator-heron/issues. The first Apache Heron (Incubating) release was voted on and passed in November 2018, and all releases are available at https://github.com/apache/incubator-heron/releases.

The Heron Slack channel is located at https://heronstreaming.slack.com/. You can self-register at http://heronstreaming.herokuapp.com/. Do not forget to follow @heronstreaming on Twitter!

1.4 Stream Processing Tools

Besides Heron, the often seen stream processing tools include Apache Flink and Spark Streaming.

Table 1.2 Stream processing tools

	Heron	Flink	Spark Streaming
Category	complex event processor	complex event processor	event stream processor
Event size	single	single	micro-batch
Delivery semantics	effectively-once, at-least-once	exactly-once	exactly-once, at-least-once
State management	distributed snapshot	distributed snapshot	checkpoints
Windowing	time-based, count-based	time-based, count-based	time-based
Primary abstraction	Tuple, Streamlet	DataStream	DStream
Data flow	topology	streaming dataflow	application
Latency	very low	low	medium
Auto-scaling	yes	no	yes
API	compositional, declarative	declarative	declarative
API language	Java, Python, C++, Scala	Java, Scala, SQL	Scala, Java, Python

Apache Flink   [5, 4]

Apache Flink is a framework and distributed processing engine for stateful computations over unbounded and bounded data streams. Any kind of data is produced as a stream of events. Data can be processed as unbounded or bounded streams. Unbounded streams have a start but no defined end and must be continuously processed. On the contrary, bounded streams have a defined start and end. Bounded streams can be processed by ingesting all data before performing any computations.

Flink provides multiple APIs at different levels of abstraction and offers dedicated libraries for common use cases. Three layered APIs are provided: ProcessFunction, the DataStream API, and SQL and the Table API. ProcessFunctions are for processing individual events from one or two input streams or events that are grouped in a window. Flink features two relational APIs: the Table API and SQL. Both APIs are unified APIs for batch and stream processing and leverage Apache Calcite for parsing, validation, and query optimization. Flink provides data source and sink connectors to systems such as Apache Kafka, Apache Cassandra, Elasticsearch, Apache HBase, Apache Hive, RabbitMQ, Amazon Kinesis, and Google Cloud Pub/Sub.

Spark Streaming   [8]

Spark Streaming is an extension of the core Spark API that enables scalable, high-throughput, fault-tolerant stream processing of live data streams. Data can be ingested from many sources like Kafka, Flume, Kinesis, or TCP sockets and can be processed using complex algorithms expressed with high-level functions like map, reduce, join, and window. Finally, processed data can be pushed out to file systems, databases, and live dashboards. You can apply Spark's machine learning and graph processing algorithms on data streams.

Internally, it works as follows. Spark Streaming receives live input data streams and divides the data into batches, which are then processed by the Spark engine to generate the final stream of results in batches.

Spark Streaming provides a high-level abstraction called Discretized Stream, or DStream, which represents a continuous stream of data. DStreams can be created either from input data streams from sources such as Kafka, Flume, and Kinesis or by applying high-level operations on other DStreams. Internally, a DStream is represented as a sequence of Resilient Distributed Datasets (RDDs).

The features of these tools are summarized in Table 1.2. Users may choose the best-fit tool in their particular environment according to this table.

1.5 Summary

This chapter began by discussing data processing with the Lambda and Kappa architectures and how Apache Storm is often adopted to fulfill the stream processing needs in both architectures. We then saw how Heron was motivated after several limitations were identified in Storm. Heron aims to achieve a set of core goals, and therefore Heron resources were provided. Finally, we compared stream processing tools and summarized their features.

References

1. Akram, S.W., M.Varalakshmi, J.Sudeepthi: Streaming big data analytics - current status, challenges and connection of unbounded data processing platforms. International Journal of Innovative Technology and Exploring Engineering (IJITEE) **8**(9S2), 698 (2019)
2. Cumbane, S.P., Gidófalvi, G.: Review of big data and processing frameworks for disaster response applications. ISPRS International Journal of Geo-Information **8**(9), 387 (2019)
3. Introduction to Apache Druid. https://druid.apache.org/docs/latest/design/. Visited on 2020-07
4. What is Apache Flink? — applications. https://flink.apache.org/flink-applications.html. Visited on 2020-07
5. What is Apache Flink? — architecture. https://flink.apache.org/flink-architecture.html. Visited on 2020-07
6. Heron design goals. https://heron.incubator.apache.org/docs/heron-design-goals. Visited on 2020-07
7. Kulkarni, S., Bhagat, N., Fu, M., Kedigehalli, V., Kellogg, C., Mittal, S., Patel, J.M., Ramasamy, K., Taneja, S.: Twitter heron: Stream processing at scale. In: Proceedings of the 2015 ACM SIGMOD International Conference on Management of Data, pp. 239–250 (2015)
8. Spark streaming programming guide. https://spark.apache.org/docs/latest/streaming-programming-guide.html. Visited on 2020-07
9. Apache Storm. https://storm.apache.org/. Visited on 2020-07
10. Flying faster with Twitter Heron. https://blog.twitter.com/engineering/en_us/a/2015/flying-faster-with-twitter-heron.html. Visited on 2020-07

11. Leaving the nest: Heron donated to Apache Software Foundation. https://blog.twitter.com/engineering/en_us/topics/open-source/2018/heron-donated-to-apache-software-foundation.html. Visited on 2020-07
12. Open sourcing Titter Heron. https://blog.twitter.com/engineering/en_us/topics/open-source/2016/open-sourcing-twitter-heron.html. Visited on 2020-07

Chapter 2
Heron Basics

The previous chapter introduced the Heron project background. This chapter will introduce the Heron basics, including the data model, API, concepts, architecture, modules, and the topology submission process.

The Heron system is roughly categorized into three layers, as shown in Fig. 2.1: topology, core and tools, and shared infrastructure. The (upper) topology layer is used by topology writers to describe business logic with a spout-and-bolt DAG. The topology deliverables are JARs or tape archives (TARs), which are fed to the middle layer—core Heron layer—by Heron command-line interface (CLI) tools. The (middle) core and tools layer runs the Heron processes to fulfill topology on top of a set of containers through a service provider interface (SPI). The (bottom) shared infrastructure layer includes the container resource pool and the service discovery service.

In this chapter, we will introduce the topology layer concepts in Sect. 2.1. Then we will discuss the Heron core layer in Sect. 2.2. Next, the topology submission process will be presented in Sect. 2.3. Finally, we will run our first Heron job in Sect. 2.4.

2.1 Topology Data Model

One of the goals for Heron is to provide a Storm-compatible API. Thus, the data model for Heron and Storm is the same and they share the topology concept. A topology is a DAG. Its vertices are spouts and bolts, and its edges are streams. Spouts emit tuples into the topology, while bolts process the tuple streams and may or may not emit tuples.

Analogous to a logical query plan for a database, a topology is the logical plan for Heron. A logical plan is translated into a packing plan by some packing algorithms before actual execution. A topology developer specifies the parallelism of each

© The Author(s), under exclusive license to Springer Nature Switzerland AG 2021
H. Wu, M. Fu, *Heron Streaming*, https://doi.org/10.1007/978-3-030-60094-5_2

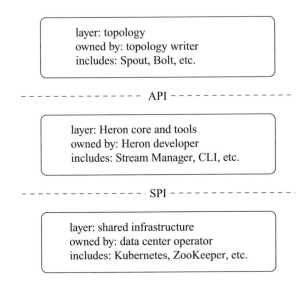

Fig. 2.1 Heron system layers

spout/bolt and grouping—how the data is partitioned between spout and bolt tasks. During actual execution, the physical device information is bonded to the packing plan. The actual topology, the parallelism specification for each component, and the grouping specification constitute the physical execution plan that is executed on the machines [5].

Heron's tuple delivery semantics describe how many times a tuple is delivered and include the following [2]:

At-most-once
    No tuple is processed more than once, although some tuples may be dropped and thus miss being analyzed by the topology.
At-least-once
    Each tuple is guaranteed to be processed at least once, although some tuples may be processed and taken into account by the topology more than once and may contribute to the result of the topology multiple times.
Effectively-once
    Each tuple is processed once. The tuples may be processed multiple times when failures happen; however, the final results look the same as the tuple is processed once and only once.

2.1.1 Topology

A Heron topology is a DAG for processing streams of data, which consists of two primary components: spouts and bolts connected by tuple streams. Figure 2.2 is an example topology. Spouts feed tuples into the DAG, while bolts process those tuples.

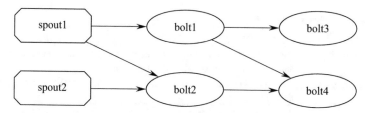

Fig. 2.2 Heron topology

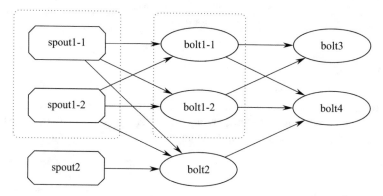

Fig. 2.3 Topology with parallelism

In Fig. 2.2, *spout1* feeds tuples to *bolt1* and *bolt2* for processing, while *spout2* feeds tuples to only *bolt2*; *bolt1* feeds processed tuples to *bolt3* and *bolt4*, while *bolt2* feeds processed tuples to only *bolt4*. You can wire bolts and spouts to compose complex DAGs.

The nodes in a topology are usually run as multiple instances in parallel. Figure 2.3 shows the same topology with parallelism for Fig. 2.2. The parallelism of *spout1* and *bolt1* is two, while the parallelism of the remaining nodes is one.

2.1.2 Spout

A spout acts as a source of streams, responsible for injecting data into Heron topologies. A spout may either generate streams itself or read data from external sources, for example, from a Kafka queue or the Google Cloud Pub/Sub, and emit tuples to bolts.

Table 2.1 Grouping strategies

Grouping strategy	Description
Fields grouping	Tuples are transmitted to bolts based on a given field. Tuples with the same field will always go to the same bolt.
Global grouping	All tuples are transmitted to a single instance of a bolt with the lowest task ID.
Shuffle grouping	Tuples are randomly transmitted to different instances of a bolt.
None grouping	Currently, this is the same as the shuffle grouping.
All grouping	All tuples are transmitted to all instances of a bolt.
Custom grouping	User-defined grouping strategy.

2.1.3 Bolt

A Heron bolt applies user-defined processing logic to data supplied by spouts, which may include transforming streams, aggregating streams, storing outputs to external storages, and sending tuple streams to other bolts.

2.1.4 Grouping

In addition to the component specification, you also need to specify how tuples will be routed between your topology components. There are a few different grouping strategies available, as shown in Table 2.1 [4].

2.2 Heron Architecture and Components

The design goals of Heron include reducing the workload of maintenance, increasing developer productivity, and improving performance predictability. The multi-objectives require careful design decisions on various system components, keeping concise abstractions and scalable architecture in mind. The Heron architecture is shown in Fig. 2.4. Users write topologies with the Heron API and deploy topologies to a resource pool using the Heron CLI. Kubernetes is an often-used scheduler for Heron to manage container pools. However, the Heron architecture abstracts a scheduler interface that makes running Heron on other schedulers easy. Heron chose to work with mature open-source schedulers, such as Marathon, YARN, and Kubernetes, instead of implementing another Heron-specific scheduler.

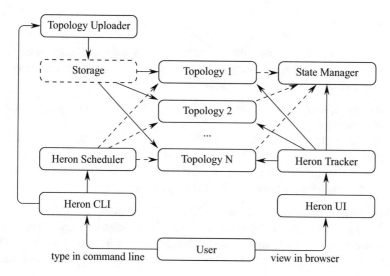

Fig. 2.4 Heron architecture

2.2.1   Cluster-Level Components (Six Components)

Heron runs in clusters driven by Heron Scheduler. The cluster could be a multi-tenant cluster or a dedicated cluster. Multiple clusters can coexist in the data centers, and a user can manage topologies in any of these clusters isolatedly. Figure 2.4 shows an often seen Heron deployment in a data center.

Several components are required to work together for Heron deployment. The following six components must be deployed to run Heron topologies in a cluster [1].

2.2.1.1   Scheduler

Heron requires a scheduler to run its topologies. It can be deployed on an existing cluster running alongside other big data frameworks. Alternatively, it can be deployed on a cluster of its own. Heron currently supports several scheduler options: Aurora, Kubernetes, Marathon, Mesos, Nomad, Slurm, YARN, and Local.

2.2.1.2   State Manager

Heron State Manager (statemgr) tracks the state of all deployed topologies. The topology state includes its logical plan, physical plan, and execution state. Heron supports the following State Managers: ZooKeeper and LocalFileSystem.

2.2.1.3 Uploader

The Heron uploader and downloader pair distributes the topology JARs to the containers that run them. Topology Uploader and Topology Downloader may work in several scenarios due to different schedulers:

- Topology Uploader copies the JAR files directly to the containers, such as the local scheduler.
- Topology Uploader uploads the JAR files to some storage from where the topology downloads the files to its containers by Topology Downloader.
- For some schedulers, Topology Uploader is not necessary; instead, the scheduler already provides the uploader and downloader functions.

Heron supports several uploaders: HDFS, DistributedLog, HTTP, SCP, S3, Google Cloud Storage, LocalFileSystem, and NullUploader. They fall into three categories: cloud services (S3, Google Cloud Storage), remote services (HDFS, DistributedLog, HTTP, SCP, NullUploader), and local services (LocalFileSystem).

2.2.1.4 Heron CLI

Heron provides a CLI tool called `heron` that is used to manage topologies. More details can be found in Chap. 8.

2.2.1.5 Heron Tracker

Heron Tracker serves as a gateway to explore the topologies. It exposes a REST API for exploring the logical plan and physical plan of the topologies and also for fetching metrics from them.

2.2.1.6 Heron UI

Heron UI provides the ability to find and explore topologies visually. It displays the DAG of the topology and how the DAG is mapped to physical containers running in clusters. Furthermore, it allows users to view logs, take a heap dump, view memory histograms, show metrics, etc.

2.2.2 Topology-Level Components (Four Components)

Each topology is run in several containers as a job, as shown in Fig. 2.5. The containers fall into two categories—the controller container and processing containers. The first container is the controller container, which runs a Topology Master

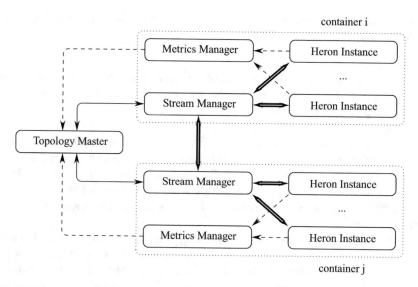

Fig. 2.5 Heron topology architecture

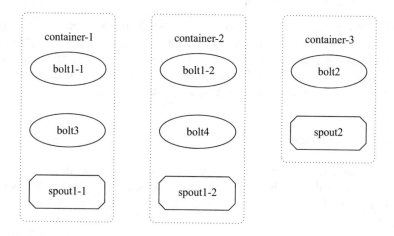

Fig. 2.6 Packing plan

process, a MetricsCache Manager process, and a Health Manager process. The rest of the containers are processing containers, each of which hosts a Stream Manager process, a Metrics Manager process, and several Heron Instance (for corresponding spouts and bolts) processes. A single physical node can host multiple containers. These containers are managed by Heron Scheduler and grouped into clusters. The topology metadata, such as the job submitter, the launch time, and the execution state, are kept in State Manager—usually ZooKeeper. Protocol buffers (protobufs) packets on top of TCP connections carry all communication between

Heron processes. To put topology components into containers, three plans are involved in the Heron life cycle [3]:

Logical plan
  A logical plan is analogous to a database query plan. Figures 2.2 and 2.3 are examples of logical plans. The topology writer is responsible for describing the logical plan with Heron APIs.
Packing plan
  Heron runs on top of a container pool. The processes in the logical plan have to be packed into a set of containers, resulting in a packing plan. Figure 2.6 shows a packing plan for Fig. 2.3. When the --dry-run option is fed to the Heron CLI, it prints the packing plan, which we will see in Chap. 8. Different packing algorithms may generate different packing plans for a given logical plan.
Physical plan
  A physical plan is derived from the packing plan; however, a physical plan adds the actual execution contexts, for example, the spout or bolt host machines, the Stream Manager process meta information, and the Metrics Manager process meta information. A packing plan may be deployed multiple times, each time leading to a different physical plan.

The four topology-level components are Heron Instance, Stream Manager, Topology Master, and Metrics Manager. Heron Instance and Stream Manager construct the data plane; Stream Manager and Topology Master construct the control plane; Metrics Manager and Topology Master construct the metric plane. The following sections describe the four components in detail [1]:

2.2.2.1 Heron Instance

Heron Instance (HI or instance for short) is a process that handles a single task of a spout or bolt, which allows for easy debugging and profiling. Currently, Heron supports Java, Python, and C++ HIs corresponding to Heron low-level API languages.

2.2.2.2 Stream Manager

Stream Manager (SM or stmgr for short) manages the routing of tuples between topology components. Each HI in a topology connects to its local SM, while all SMs in a given topology connect to form an SM network. There are three possible SM connection patterns: fully distributed, centralized, and decentralized.

Fully distributed communication
  The communication can be distributed entirely so that each HI/SM connects to all the other HI/SMs to form a complete graph. Assume there are N HI/SMs;

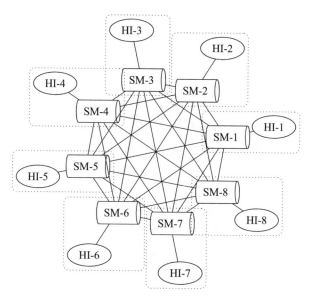

Fig. 2.7 Completely distributed communication

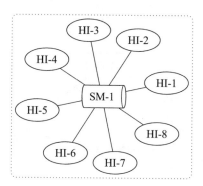

Fig. 2.8 Central communication bus

then, there are $N(N-1)/2$ edges. For a topology with eight HIs, its simplified physical plan may look like Fig. 2.7.

Central communication bus

There are too many edges in a wholly distributed communication pattern. The opposite is adopting a superpower central communication node connecting to all HIs, which is responsible for exchanging packets like a central network switch. The simplified physical plan looks like Fig. 2.8.

Semi-centralized, semi-distributed architecture

This fully centralized network in Fig. 2.8 may be unstable when scaling up, and it is hard for the central exchanging node to achieve high throughput. Heron picks the semi-centralized, semi-distributed architecture consisting of M nodes

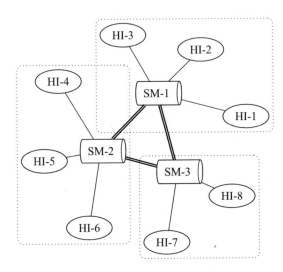

Fig. 2.9 Semi-centralized, semi-distributed architecture

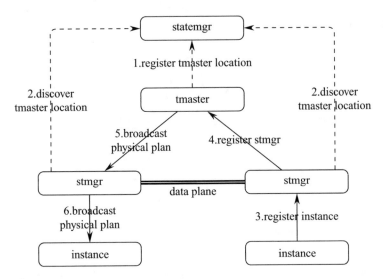

Fig. 2.10 Registration between TM, SM, and HI

corresponding to M containers, where $1 < M < N$. The simplified physical plan looks like Fig. 2.9 for the example packing plan in Fig. 2.6.

2.2.2.3   Topology Master

Topology Master (TM or tmaster for short) manages a topology throughout its entire life cycle, from the time it is submitted until it is ultimately killed. When Heron

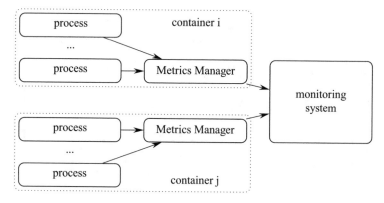

Fig. 2.11 Metrics flow

deploys a topology, it starts a single TM and multiple containers. The TM creates an ephemeral State Manager node to ensure that there is only one TM for the topology and that the TM is easily discoverable by any process in the topology. The TM also constructs the physical plan for a topology, which it relays to the different components shown in Fig. 2.10.

2.2.2.4 Metrics Manager

Each topology runs a Metrics Manager (MM or mm for short) that collects and exports metrics from all components in each container shown in Fig. 2.11. It then routes those metrics to both internal collectors, such as Topology Master and MetricsCache Manager, and external collectors, such as Scribe, Graphite, or analogous systems. You can adapt Heron to support new monitoring systems by implementing your custom metrics sink.

2.3 Submission Process and Failure Handling

When a user submits a topology to Heron, the following steps are triggered (see Figs. 2.12 and 2.13). Upon submission, the Heron CLI uploads the topology package to shared storage that is accessible from the containers. The Heron CLI also updates the ZooKeeper metadata and triggers Heron Scheduler to allocate the required resources and schedule the containers in a cluster. The containers come up and identify their role in the cluster—either controller container or processing container. The controller container spawns the TM process that registers itself as an ephemeral ZooKeeper node. At the same time, the processing containers spawn the SM processes, which connect to ZooKeeper and discover the TM location. Once the TM location is available, the SM connects to the TM and starts sending

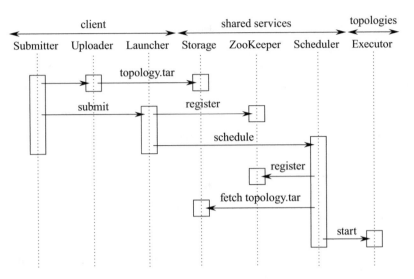

Fig. 2.12 Topology submission sequence

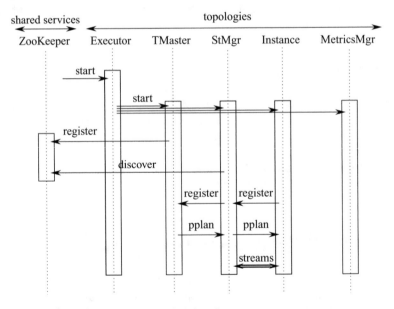

Fig. 2.13 Topology submission sequence (continued)

heartbeats. After all the SMs connect to the TM, the TM assigns containers in the packing plan to the actual containers that the SM represents. This mapping from the packing plan to the actual containers is called a physical plan. Once the physical plan is complete, the TM broadcasts it to the SMs; thus, the SMs can discover each other and connect to form a fully connected graph. Meanwhile, HIs come up in the processing containers, connect to the local SM, fetch the physical plan, and start running the assigned spout or bolt user logic code. Once these steps are completed, the topology is ready for tuples to flow through. For safekeeping, the TM writes the physical plan to ZooKeeper to rediscover the state in case of failure [5].

For a job running at scale, failures are inevitable. For Heron, these failures could be process death, container failure, and machine failure [5]:

Process death

If the TM process dies, Heron Executor restarts the failed TM process, and the new TM registers itself to ZooKeeper and recovers its state. Meanwhile, the SMs monitoring ZooKeeper find the new TM and connect to it.

Similarly, if an SM dies, Heron Executor restarts it in the same container, and the new SM discovers the TM and connects to the TM fetching the physical plan to update its state. Other SMs get a copy of the same new physical plan, including the new SM location, and establish a connection with the new SM.

When an HI process dies in a container, it is restarted and contacts its local SM. The SM sends the HI a copy of the physical plan indicating its identity and starts processing the corresponding user logic code.

Container and machine failures

If the container orchestrator reschedules any container to a new machine, the new container follows the same steps of an SM failure and an HI failure. If any machine fails, the containers on that machine should be rescheduled by the container orchestrator, and the new containers follow the previous failure handling steps.

2.4 Submit the First Topology

After introducing the theory behind Heron, this section will demonstrate how to submit a simple example topology in local mode and observe its running state. This simple demonstration will provide readers a taste of Heron, while Chaps. 3 and 8 will cover this in more detail.

Although Heron can run on both Linux and macOS, we use the environment in Fig. 1 for the examples in this book. To run the examples on macOS, you have to use the corresponding commands in macOS. Moreover, the examples used in this book run on a single machine for demonstration purposes. Heron is highly configurable and runs at scale in data centers.

2.4.1 Preparation

Install language-compiling tools for Java, Python, and C++. Install the project management tool Bazel.

```
$ sudo apt update  ①

$ sudo apt install -y openjdk-11-jdk && javac -version  ②
$ sudo apt install -y python-is-python3 python3-dev && \
> python --version  ③
$ sudo apt install -y \
> build-essential automake libtool-bin ant libcppunit-dev  ④

$ sudo apt-get -y install pkg-config zip unzip tree && \
> wget -O /tmp/bazel-3.4.1-installer-linux-x86_64.sh \
> https://github.com/bazelbuild/bazel/releases/download\
> /3.4.1/bazel-3.4.1-installer-linux-x86_64.sh && \
> chmod +x /tmp/bazel-3.4.1-installer-linux-x86_64.sh && \
> /tmp/bazel-3.4.1-installer-linux-x86_64.sh --user && \
> export PATH="$PATH:$HOME/bin" && \
> bazel version  ⑤
```

① Update the APT repository.
② Install JDK11 for Java. See Sect. 3.1 for details on the supported languages and their versions.
③ Install Python3 for Python.
④ Install the default toolchain for C++.
⑤ Install Bazel version 3.4.1.

Download the Apache Heron (Incubating) version 0.20.3-incubating-rc7 source code from the Apache website.[1] Then verify that the compiling dependency is satisfied by running a script shipped with the source code.

```
$ cd ~ && \
> wget https://dist.apache.org/repos/dist/dev/incubator/heron\
> /heron-0.20.3-incubating-candidate-7\
> /0.20.3-incubating-rc7.tar.gz && \
> tar -xzf 0.20.3-incubating-rc7.tar.gz && \
> mv incubator-heron-0.20.3-incubating-rc7/ heron/ && cd heron/

$ ./bazel_configure.py  ①
```

① This script prints the detected compiling environment. If any dependency is not met, it prints errors.

---

[1]The release candidates are hosted at https://dist.apache.org/repos/dist/dev/incubator/heron/, while the releases are hosted at https://dist.apache.org/repos/dist/release/incubator/heron/ as well as https://downloads.apache.org/incubator/heron/.

2.4.2   Install the Heron Client

Compile a Heron client binary and install it under the current user role:

```
$ cd ~/heron/ && bazel build \
> --config=ubuntu_nostyle scripts/packages:binpkgs  ①

$ ./bazel-bin/scripts/packages/heron-install.sh --user  ②
$ ~/.heron/bin/heron version  ③
```

①   The compile process takes time, usually half an hour.
②   Run the Heron CLI installation script, which extracts the Heron CLI and
       installs it to the current user home directory.
③   Test the `heron` command, which should be available in the hidden directory
       ~/.heron/bin/, to verify that the installation is successful.

NOTE

The Heron CLI requires JAVA_HOME. You can set the path by:

```
$ export JAVA_HOME=/usr/lib/jvm/java-11-openjdk-amd64
```

If the command or the log complains that the local machine hostname cannot be
resolved, add it to /etc/hosts:

```
$ sudo cp /etc/hosts /etc/hosts.bak && \
> echo "127.0.0.1 `hostname`" | sudo tee --append /etc/hosts
```

2.4.3   Heron Example Topologies

Along with the Heron CLI, several example topologies are installed:

```
$ ls -l ~/.heron/examples/*.jar  ①
heron-api-examples.jar  ②
heron-eco-examples.jar  ③
heron-streamlet-examples.jar  ④
heron-streamlet-scala-examples.jar  ⑤
storm-eco-examples.jar  ⑥
```

①   List the installed example topology files.
②   Topologies composed by the low-level Topology API. We will see the low-
       level Topology API in Chap. 5.
③   Components to compose topologies by the Extensible Component Orches-
       trator (ECO).
④   Processing graphs composed by the high-level Streamlet API. We will see
       the high-level API in Chap. 7.

⑤   Processing graphs composed by the Scala Streamlet API.
⑥   Components composed by the compatible Storm API for the ECO.

Among these example topologies, the `ExclamationTopology` topology in *heron-api-examples.jar* is the most often seen basic topology.

2.4.4   Submit the Topology JAR File

Submit the `ExclamationTopology` topology with the name *hello-world-topology*:

```
$ ~/.heron/bin/heron submit local \①
> ~/.heron/examples/heron-api-examples.jar \②
> org.apache.heron.examples.api.ExclamationTopology \③
> hello-world-topology ④
```

①   Submit the topology to the local host.
②   The JAR file including the topology definition.
③   The Java classpath indicating the `main()` entry.
④   A string to name the first topology. The same string is used to locate the same topology in the kill command.

When the topology is submitted successfully, the Heron CLI prints:

```
[INFO]: Successfully launched topology 'hello-world-topology'
```

2.4.5   Observe the Running Topology

When the topology is running, a cluster of processes is boosted. A directory *~/.herondata/* is generated to host the topology state, and this directory also serves as the topologies' working directory.

```
$ PID=$(ps -ef | \
> grep SchedulerMain | grep java | awk '{print $2}')
$ [ -z "${PID}" ] && echo "job not found" || pstree -aApT ${PID}
```

This command prints the topology process tree in two steps. It first searches the processes and finds a Java process whose command arguments contain *SchedulerMain*. Then **pstree** displays the process tree according to the found process ID. This found process is the Heron Scheduler process, which we will revisit in Chaps. 10 and 12.

     To kill the topology, run:

```
$ ~/.heron/bin/heron kill local hello-world-topology
```

2.5 Summary

This chapter described the data model, including topology, spout, and bolt. Then the Heron architecture was presented and the Heron components were introduced at cluster level and topology level. Finally, the submission process was described as well as failure handling. In the last part of this chapter, you submitted your first topology. Congratulations on your first step on the Heron journey! Starting from the next chapter, we will move forward to more practical content.

References

1. Heron architecture. https://heron.incubator.apache.org/docs/heron-architecture. Visited on 2020-07
2. Heron delivery semantics. https://heron.incubator.apache.org/docs/heron-delivery-semantics. Visited on 2020-07
3. Heron topologies. https://heron.incubator.apache.org/docs/heron-topology-concepts. Visited on 2020-07
4. The Heron Topology API for Java. https://heron.incubator.apache.org/docs/topology-development-topology-api-java. Visited on 2020-07
5. Kulkarni, S., Bhagat, N., Fu, M., Kedigehalli, V., Kellogg, C., Mittal, S., Patel, J.M., Ramasamy, K., Taneja, S.: Twitter Heron: Stream processing at scale. In: Proceedings of the 2015 ACM SIGMOD International Conference on Management of Data, pp. 239–250 (2015)

Chapter 3
Study Heron Code

After Heron joined the Apache family, the source code is the only mandatory requirement in Heron releases. The source code is the source of truth. Once you have the Heron release source code, you can build a colorful Heron world. This chapter focuses on the Heron code and covers the following topics:

- How to get ready for compiling the Heron source code in Sects. 3.2 and 3.3
- How the Heron code is organized in Sects. 3.1 and 3.4
- How to compile the Heron code in Sect. 3.5 and what outputs are expected in Sect. 3.6
- How to run the tests in the Heron code in Sect. 3.7

3.1 Code Languages

Heron is made primarily with Java, C++, and Python programming languages:

Java 11   is used primarily for Heron's core component, Heron Instance, and metrics-related components, including Metrics Manager, MetricsCache Manager, and Health Manager.

C++ 11   is used for traffic relaying nodes and the topology controller, including Stream Manager and Topology Master.

Python 3   is used primarily for tools, including the Heron CLI, Heron Tracker, Heron UI, and Heron Explorer. Heron Executor is also in Python, which spawns other processes inside the container.

Besides the above primary programming languages, Heron also uses JavaScript, Starlark, Shell, and others in different parts. The code stats from GitHub are shown in Fig. 3.1.

© The Author(s), under exclusive license to Springer Nature Switzerland AG 2021
H. Wu, M. Fu, *Heron Streaming*, https://doi.org/10.1007/978-3-030-60094-5_3

Languages

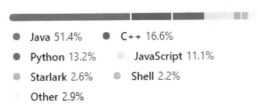

- ● Java 51.4%   ● C++ 16.6%
- ● Python 13.2%   ● JavaScript 11.1%
- ● Starlark 2.6%   ● Shell 2.2%
- ● Other 2.9%

Fig. 3.1 Code languages (July 2020)

Table 3.1 Requirements for compiling

Tool	Version	Notes
Bazel[1]	= 3.4.1	Heron uses Bazel as its build tool. Later versions might work but have not been tested.
Java	>= 11	Bazel and Heron require Java 11. Topologies can be written in Java 8 or above, but all Heron JARs are required to run with a Java 11 JRE.
Python	>= 3.6	
gcc/g++[2]	>= 4.8.1	
Autoconf[3]	>= 2.6.3	Used for third-party dependencies, such as ZooKeeper client library.

[1] http://bazel.io/docs/install.html
[2] https://gcc.gnu.org/
[3] http://www.gnu.org/software/autoconf/autoconf.html

NOTE

The topology code languages are slightly different from the Heron code languages. They are determined by the languages of Heron Instance, which is further determined by the protocol buffers languages. The often-used programming languages for topology are Java and Python.

Java 8   or later is used for Heron's Topology API. We will write topology code in Sect. 5.2.1. The Java API document can be found at https://heron. incubator.apache.org/api/java/.

Python 3   is also used for Heron's Topology API. We will try an example Python topology in Sect. 5.3. The Python API document can be found at https:// heron.incubator.apache.org/api/python/.

3.2 Requirements for Compiling

In this book, we aim at compiling Heron in the Linux environment. The compile requirements are listed in Table 3.1. These requirements have to be satisfied to compile Heron 0.20.3-incubating-rc7.

3.3   Prepare the Compiling Environment

From this point on in the book, we begin to work in the experiment environment described in Fig. 1. Before compiling the Heron source code, we have to meet the requirements in Table 3.1. The following steps show how to prepare a fresh Ubuntu Server 20.04 for compiling Heron:

1. Update the package list and upgrade the system.

```
$ sudo apt update && sudo apt full-upgrade -y
```

2. Install JDK 11.

```
$ sudo apt install -y openjdk-11-jdk && javac -version
$ export JAVA_HOME=/usr/lib/jvm/java-11-openjdk-amd64
```

3. Ubuntu 20.04 includes Python 3 rather than Python 2 by default. Make sure it is available:

```
$ python3 -V
$ sudo apt install -y python-is-python3
$ python -V
```

4. The Bazel installation guide is available at https://docs.bazel.build/versions/master/install-ubuntu.html. To compile the Heron 0.20.3-incubating-rc7 source code, Bazel version 3.4.1 is used. Install Bazel:

```
$ sudo apt-get -y install unzip && \
> wget -O /tmp/bazel-3.4.1-installer-linux-x86_64.sh \
> https://github.com/bazelbuild/bazel/releases/download\
> /3.4.1/bazel-3.4.1-installer-linux-x86_64.sh && \
> chmod +x /tmp/bazel-3.4.1-installer-linux-x86_64.sh && \
> /tmp/bazel-3.4.1-installer-linux-x86_64.sh --user && \
> export PATH="$PATH:$HOME/bin" && \
> bazel version
```

5. Install the required libraries and tools.

```
$ sudo apt install -y \
> build-essential automake libtool-bin \ ①
> ant libcppunit-dev pkg-config \ ②
> python3-dev tree \ ③
> zip ④
```

①	Tools for building C++ targets. Required by the compiling environment validation check *bazel_confgure.py* in Sect. 3.5.
②	Dependencies for building ZooKeeper C++ targets.
③	Python-related dependencies.
④	Tools for generating documents.

3.4   Source Organization

In Sect. 2.4, we saw how to download a specific release version source code from
the Apache repository, which is used for this book. The Heron source code is hosted
at the GitHub website,[1]   rather than the Apache website. Alternatively, we can
download the source from GitHub and pick the target version:

```
$ cd ~ && wget https://dist.apache.org/repos/dist/dev\
> /incubator/heron/heron-0.20.3-incubating-candidate-7\
> /0.20.3-incubating-rc7.tar.gz && \  ①
> tar -xzf 0.20.3-incubating-rc7.tar.gz && \
> mv incubator-heron-0.20.3-incubating-rc7/ heron/ && cd heron/

$ cd ~ && git clone \
> https://github.com/apache/incubator-heron.git && \  ②
> mv incubator-heron/ heron/ && cd heron/ && \
> git checkout tags/0.20.3-incubating-rc7
```

①   Download the source code from the Apache website, which is used in this
book.
②   Alternatively, download the source code from the GitHub repository.

3.4.1   Directory Organization

Before compiling the source code, let us first browse the source code directory to
see how the source code is organized. The major directories and files are:

```
$ ls -l ~/heron/
WORKSPACE ①
bazel_configure.py ②
config ③
deploy ④
docker ⑤
eco ⑥
eco-heron-examples ⑦
eco-storm-examples ⑧
examples ⑨
heron ⑩
heronpy ⑪
integration_test ⑫
maven_install.json ⑬
release ⑭
scripts ⑮
storm-compatibility ⑯
storm-compatibility-examples ⑰
third_party ⑱
tools ⑲
```

---

[1]https://github.com/apache/incubator-heron.

①   Bazel requires this project file.
②   A Python script to validate the compiling dependencies and configure Bazel on supported platforms.
③   The configurations required during the compiling process, especially for C++-related Bazel targets.
④   Tools for running the Heron Docker and Kubernetes clusters.
⑤   Tools for compiling Heron in a Docker container and for packaging a Heron Docker image.
⑥   The ECO code directory, which is similar to Storm Flux. It helps the topology writer use the YAML file to wire spouts and bolts into a topology.
⑦   The ECO examples based on the Heron API.
⑧   The ECO examples based on the Storm API. Since the Heron API is compatible with the Storm API, the ECO can generate topology through either the Heron API or the Storm API.
⑨   A set of example topologies built using *heron/api/* and *heronpy/*. The example topologies are written in Java, Python, Scala, and C++.
⑩   The primary Heron code directory, including all codes for API/SPI, common, components, tools, etc. This directory is where most code exists. We will examine its content soon.
⑪   The Heron API for Python, which is a part of *heron/api/*. The API documentation can be found at https://heron.incubator.apache.org/api/python/.
⑫   Integration test code. There are three sets of integration tests, including topology correctness tests of the Java, Python, and Scala topologies, and topology structure tests.
⑬   The Bazel cache file for the Maven dependencies.
⑭   Tools for publishing a Heron release.
⑮   Miscellaneous scripts, including CI scripts, integration test scripts, and release scripts.
⑯   Wrap *heron/api/* to implement the Apache Storm API.
⑰   Topology examples based on the Apache Storm API.
⑱   The Heron dependencies.
⑲   Bazel-related files; used by Bazel.

The major source code exists in the *heron/* directory. Let us have a close look at it:

```
$ ls -l ~/heron/heron/
api ①
ckptmgr ②
common ③
config ④
downloaders ⑤
executor ⑥
healthmgr ⑦
instance ⑧
io
metricscachemgr ⑨
metricsmgr ⑩
```

```
packing  ⑪
proto  ⑫
scheduler-core  ⑬
schedulers  ⑭
shell  ⑮
simulator  ⑯
spi  ⑰
statefulstorages  ⑱
statemgrs  ⑲
stmgr  ⑳
tmaster  ㉑
tools  ㉒
uploaders  ㉓
```

① The Heron API for Java, which is used by topology writers to write topologies. The API documentation can be found at https://heron.incubator. apache.org/api/java/.

② The Checkpoint Manager code. The Checkpoint Manager process runs in the non-zero containers when the topology reliability mode is set to `effectively-once`.

③ A variety of utilities for each code language, including useful constants, Heron servers and clients, file/shell utilities, networking/backpressure utilities, and more.

④ YAML configuration files, most of which are used by *heron/spi/* implementations.

⑤ The Topology Downloader code. Topology Downloader is responsible for downloading any packages to the local container, especially to download the topology *TAR.GZ* file, including the topology JAR, configuration directory, and topology definition file.

⑥ The Heron Executor code. Heron Executor is the first process launched inside containers. It monitors the other Heron processes inside the container.

⑦ The Health Manager code. Health Manager runs inside container 0 and is responsible for auto-healing the topology.

⑧ The source code for Heron Instance, which is the actual processor for spouts and bolts.

⑨ MetricsCache Manager runs inside container 0 to cache the metrics. It replaces the metrics collector in Topology Master so that Topology Master focuses more on topology management tasks.

⑩ The source code for Metrics Manager and metrics handlers (known as metrics sinks) in Java.

⑪ The directory for packing algorithm implementations. The packing algorithm calculates how to pack processes into the containers.

⑫ Heron uses protocol buffers for packet serialization/deserialization. This directory hosts most **.proto* definition files.

⑬ The common Heron Scheduler functions, including submitter, launcher, runtime management, etc.

⑭   This directory hosts codes for each particular Heron Scheduler. Heron
        supports several cluster schedulers: Kubernetes, YARN, Mesos, and a
        local scheduler. You can also implement your schedulers. All the Heron
        Schedulers implement SPI in *heron/spi/*.
⑮   Heron Shell code in Python. Heron Shell is responsible for container status
        queries.
⑯   Since Heron processes are distributed, and it is not convenient to debug
        multiple processes at the same time, `Simulator` was introduced to run the
        processes as threads in a single JVM process for debuggers to attach to.
⑰   The Heron SPI for Java used to extend Heron by contributors.
⑱   When "effectively-once" is set, a state store is mandatory.
⑲   The State Manager directory includes two often-used implementations:
        ZooKeeper and the local file system. ZooKeeper is for Heron distributed
        deployment, while the local file system is for local mode.
⑳   The source code for Stream Manager in C++.
㉑   The source code for Topology Master in C++.
㉒   Code for Heron tools, including Heron client, Heron Tracker, and Heron UI.
㉓   Topology Uploader is responsible for uploading the topology *TAR.GZ* file to
        a shared remote store for containers to fetch during the topology launching
        process.

3.4.2   Bazel Perspective

Bazel manages the Heron source code. It is an open-source software tool for the
automation of building and testing software. Google internally uses the build tool
Blaze and released part of the Blaze tool as Bazel—an anagram of Blaze. This
section discusses how code is organized from the Bazel perspective.

Bazel builds software from the source code organized in a directory called a
workspace. Each workspace directory has a text file named *WORKSPACE*. Source
files in the workspace are organized in a nested hierarchy of packages, where each
package is a directory that contains a set of related source files and one *BUILD* file.
The *BUILD* file specifies what software outputs can be built from the source [1]. We
may check how many *BUILD* files there are in Heron:

```
$ cd ~/heron/ && bazel query ... --output package
```

A package is a container. The elements of a package are called targets. Most
targets are one of two principal kinds: files and rules. Files are further divided
into two kinds: source files and generated files (sometimes called derived files).
A rule specifies the relationship between a set of input and output files, including
the necessary steps to derive the outputs from the inputs. The outputs of a rule are
always generated files [1]. The following command tells how many targets are inside

these *BUILD* files, as well as the label and kind of each target:

```
$ bazel query ... --output label_kind
```

Particularly, we may process the display to show the unique rules (the output is organized by categories):

```
$ bazel query ... --output label_kind | cut -d' ' -f1 | sort -u
jarjar_binary  ①
java_binary
java_doc
java_library
java_test

cc_binary  ②
cc_library
cc_test
cc_toolchain
cc_toolchain_config

pex_binary  ③
pex_library
_pytest_pex_test

scala_binary  ④
scala_library
scala_test

self_extract_binary  ⑤
sh_binary
sh_library

genproto_java  ⑥
proto_package

container_image_  ⑦
container_push

action_listener  ⑧
extra_action

config_setting  ⑨
filegroup
genrule
pkg_tar_impl
```

①	Java-related rules.
②	C++-related rules.
③	Python-related rules.
④	Scala-related rules.
⑤	Shell-related rules.
⑥	Protocol buffers rule to generate Java code.
⑦	Docker container-related rules.

⑧ Rules to check code styles.
⑨ Common rules.

We may query what *BUILD* files are involved in building a particular target with the following command (`proto` for example):

```
$ bazel query 'buildfiles(//heron/proto/...)'
@rules_java//...
@rules_cc//...
@io_bazel_rules_scala//...
@io_bazel_rules_docker//...
@bazel_tools//...
@bazel_skylib//...
//tools/...
//heron/proto:BUILD
```

3.5 Compile Heron

Heron can be built as an installation script or as a series of TAR packages as a whole. Each particular Heron component, such as the Java and Python APIs, can also be built separately. We are going to build an installation script and API packages that will be used in later chapters.

1. Validate and configure the compiling dependencies for building with Bazel.

   ```
   $ cd ~/heron/ && ./bazel_configure.py
   ```

2. Build all targets in *BUILD* files in the *heron/* directory.

   ```
   $ bazel build --config=ubuntu_nostyle heron/...
   ```

   This step may take a long time. It runs for half an hour on the authors' Ubuntu machine.

3. Build the packages.

   The following **bazel build** commands build bundled TARs, an installation script, and the Heron Python API package, respectively.

   ```
   $ bazel build --config=ubuntu_nostyle \
   > scripts/packages:tarpkgs ①
   $ bazel build --config=ubuntu_nostyle \
   > scripts/packages:binpkgs ②
   $ bazel build --config=ubuntu_nostyle \
   > scripts/packages:pypkgs ③
   ```

   ① Build three Heron *TAR.GZ* files: *heron-core.tar.gz*, *heron-tools.tar.gz*, and *heron.tar.gz*. These *TAR.GZ* files are in the *bazel-bin/scripts/packages/* directory.

   ② Build an installation script *bazel-bin/scripts/packages/heron-install.sh* to install the Heron CLI. The Heron CLI package has the same content as *heron.tar.gz*.

   ③ Build the Python API package in *bazel-bin/scripts/packages/heronpy.whl*.

NOTE: Bazel OS Environments

Bazel builds are specific to a given platform. A platform-specific configuration should be indicated using the -**config** flag. The following OS values are supported: *ubuntu* (Ubuntu), *debian* (Debian), *centos* (CentOS), and *darwin* (Mac OS X), which are defined in the *tools/bazel.rc* file. The corresponding *_nostyle* configurations remove code style checks to save compiling time.

NOTE: Building Specific Components

If you are interested in a particular Heron component, you can build it individually by specifying the **bazel build** target. For example, the following command builds Heron Tracker:

```
$ bazel build --config=ubuntu_nostyle \
> heron/tools/tracker/src/python:heron-tracker
```

After Heron Tracker is built successfully, the executable is located at *bazel-bin/heron/tools/tracker/src/python/heron-tracker*.

TIP: Enabling Optimizations

Performance optimizations for C/C++ codes are not enabled by default but required by production release. To enable optimization and remove debugging information, add the **opt** flag:

```
$ bazel build -c opt --config=ubuntu_nostyle \
> heron/tools/tracker/src/python:heron-tracker
```

3.6 Examine Compiling Results

The **bazel build** command generates several directories in the Heron source root directory:

```
$ ls -l ~/heron/bazel-*
bazel-heron       -> /.../org_apache_heron ①
bazel-out         -> /.../.../bazel-out ②
bazel-bin         -> /.../.../bazel-out/k8-fastbuild/bin ③
bazel-testlogs    -> /.../.../bazel-out/k8-fastbuild/testlogs ④
```

① Working tree for the Bazel build and root of symlink forest: *execRoot*.
② All actual outputs of the build.

③   Bazel outputs binaries for target configuration.
④   Bazel internal test runner puts test log files here.

3.6.1 Examine the API

The Java API JAR and Python API packages are generated by the command in the previous section. They are located at:

```
$ ls -l bazel-bin/heron/api/src/java/*.jar
bazel-bin/heron/api/src/java/heron-api.jar ①
$ ls -l bazel-bin/scripts/packages/*.whl
bazel-bin/scripts/packages/heronpy.whl ②
```

①   The Heron API JAR file, which is supposed to be uploaded to the Maven repository and which will be used in later chapters to build Java topologies.
②   The Heron API wheel file, which is supposed to be uploaded to the PyPI repository and which will be used in Sect. 5.3.

3.6.2 Examine Packages

Target `scripts/packages:tarpkgs` generates three bundled *TAR*s in the *bazel-bin/scripts/packages/* directory: *heron-core.tar.gz*, which should be downloaded to the container during the topology launching process; *heron-tools.tar.gz*, which includes all command-line tools and their configuration YAML files; and *heron.tar.gz*, which is the Heron client, including not only all content in *heron-core.tar.gz* and *heron-tools.tar.gz* but also example topologies and the C++ API.

Target `scripts/packages:binpkgs` generates one self-extracting executable installation script, which will be used in later chapters:

```
$ ls -l bazel-bin/scripts/packages/*.sh
bazel-bin/scripts/packages/heron-install.sh ①
```

①   Used in Parts II and III.

3.7 Run Tests

To verify the code is complete and correct, run tests. Heron includes various tests, including unit tests and integration tests scattered throughout the codebase.

3.7.1 Unit Test

You can build and run unit tests using Bazel. The unit tests are mainly located in the
heron/ directory. The following command will run all test targets in *heron/*:

```
$ bazel test --config=ubuntu_nostyle heron/...
```

You can also run a particular unit test with its Bazel target name, for example:

```
$ bazel test --config=ubuntu_nostyle \
> heron/statemgrs/tests/java:LocalFileSystemStateManagerTest
```

TIP: Discovering Unit Test Targets

 Although the test targets are scattered, `bazel query` can list them easily. This
command lists all Bazel test targets [2]:

```
$ bazel query 'kind(".*_test rule", ...)'
```

To list Java test targets:

```
$ bazel query 'kind("java_test rule", ...)'
```

To list C++ test targets:

```
$ bazel query 'kind("cc_test rule", ...)'
```

To list Python test targets:

```
$ bazel query 'kind("pex_test rule", ...)'
```

3.7.2 Integration Test

Integration tests run real topologies, which are categorized into three sets:

Functional integration tests
   These integration tests run a set of real topologies in Java and Python. These
   topologies are incorporated with several special spouts generating planned tuples
   and a special terminal bolt, which sends tuples to an HTTP server to compare the
   final tuples with the expected tuples. To launch the functional integration tests,
   run the following script from the source root directory:

```
$ ./scripts/run_integration_test.sh
```

Topology structure integration tests
   This topology structure testing exposes faults in the interaction between inte-
   grated units of Heron after topology operations. Topology structure tests include:
   (1) topology graph components' parallelism and graph edge correctness, and

(2) Heron Instance states in the target rollback checkpoint in stateful processing:

```
$ ./scripts/run_integration_topology_test.sh
```

Failure integration tests

The containers in the cluster are supposed to be rescheduled by the container orchestrator; thus, the Heron topology/processes should tolerate restarts. These tests emulate the failures/restarts of Heron processes and validate their recovery. To run the failure integration tests, do the following from the source root directory:

```
$ bazel build --config=ubuntu_nostyle integration_test/src/...
$ ./bazel-bin/integration_test/src/python/local_test_runner\
> /local-test-runner
```

3.8 Summary

In this chapter, we started from a fresh Ubuntu Server and installed libraries and tools to get the compiling environment ready. Then we downloaded the Heron source code from GitHub and studied how the source code is organized. Next, we verified the compiling requirements with *bazel_configure.py* and compiled the source code to get the executable scripts. We examined the compiling output and expect to use this compiling output in later chapters. Finally, we ran tests and got confident about the Heron code.

In the next chapter, we will move to the topology code. We will study how to make an existing Apache Storm topology code ready to be submitted to Heron.

References

1. Concepts and terminology. https://docs.bazel.build/versions/master/build-ref.html. Visited on 2020-07
2. Running tests. https://heron.incubator.apache.org/docs/compiling-running-tests. Visited on 2020-07

Part II
Write Heron Topologies

Chapter 4
Migrate Storm Topology to Heron

The Heron API is designed to be fully backward compatible with the existing Apache Storm topology projects, which means that you can migrate an existing Storm topology to Heron by making just a few adjustments.

In this chapter, we will first look at two ways often used by topology writers to construct Storm topologies—by Java code and by Flux YAML code. Then, we will go through one typical Storm topology for each topology composition approach. Based on the two existing examples of Storm topology codes, we will show how to make simple adjustments and get them ready to run on Heron.

The first migration example starts with the well-known Storm topology *ExclamationTopology* in Sect. 4.1.1, then modifies a bit of code in Sect. 4.2.1 and a project file in Sect. 4.2.2, and finally compiles the topology code to a JAR file ready to be submitted to Heron in Sect. 4.2.3. The second migration example starts with the Flux YAML configuration file in Sect. 4.1.2, then modifies the YAML code to the Heron ECO code, and finally submits the topology in Sect. 4.3.

4.1 Prepare the Storm Topology Code

In this section, we prepare for migration: download the Storm source, identify the Storm projects we will migrate, and examine the Storm projects that we will compare with the Heron projects.

The Apache Storm source code contains a variety of example topology projects. Two of them are adopted in this chapter: *storm-starter* and *flux-examples*. The *storm-starter* project is designed for topology writer beginners and contains several well-known topologies, among which *ExclamationTopology* is one basic topology written completely in Java. In this chapter, we will start our first migration procedures based on the *ExclamationTopology* code.

© The Author(s), under exclusive license to Springer Nature Switzerland AG 2021
H. Wu, M. Fu, *Heron Streaming*, https://doi.org/10.1007/978-3-030-60094-5_4

Storm Flux is a framework and set of utilities that make defining and deploying Storm topologies less painful and developer intensive. One of the pain points often mentioned is the fact that the wiring for a topology graph is often tied up in Java code, and that any changes require recompilation and repackaging of the topology JAR file. Flux aims to alleviate this pain by allowing you to package your entire Storm components in a single JAR and use an external text file to define the layout and configuration of your topologies [2]. The *flux-examples* project includes an example *simple_wordcount.yaml*, which is an elementary WordCount example using Java spouts and bolts. In this chapter, we will start our second migration procedures based on the *simple_wordcount.yaml* code.

Before studying the Storm code, make sure you have the Storm code and the **storm** command available on the experiment machine. The following commands show two ways of downloading the Storm code from the Apache repository and GitHub:

```
$ cd ~ && wget https://downloads.apache.org/storm\
> /apache-storm-2.2.0/apache-storm-2.2.0-src.tar.gz && \
> tar zxf apache-storm-2.2.0-src.tar.gz && \
> mv apache-storm-2.2.0/ storm/ && cd storm/  ①

$ cd ~ && git clone https://github.com/apache/storm && \
> cd ~/storm && git checkout tags/v2.2.0  ②

$ ls -l examples/storm-starter/  ③
$ ls -l flux/flux-examples/  ④
```

①   v2.2.0 is the latest Storm version as of the writing of this book. Downloading from the Apache repository is the default way to fetch the source code in this book.
②   As an alternative to the Apache repository, the source code can be downloaded from GitHub.
③   The *storm-starter* example topologies.
④   The Flux example topologies.

Once we have the Storm source code, we can build the **storm** command. The Storm project adopts Maven[1] to build the code. The **storm** command needs Java, and Flux needs Python. The following commands prepare the dependencies:

```
$ sudo apt install -y maven && mvn -version  ①

$ javac -version  ②
$ python --version  ③
```

①   Ensure the **mvn** command is available.
②   Ensure JDK 11 is available. If it is not installed, follow the steps in Sect. 3.3 to install it.

---

[1] http://maven.apache.org/.

③   Ensure Python 3 is available. If it is not installed, follow the steps in Sect. 3.3
       to install it.

Finally, build the **storm** command:

```
$ cd ~/storm/storm-dist/binary/ && \
> mvn clean package -Dgpg.skip=true ①
$ cd ./final-package/target/ && \
> tar zxf apache-storm-2.2.0.tar.gz
$ cp -r ./apache-storm-2.2.0/ ~ && \
> export PATH=$PATH:~/apache-storm-2.2.0/bin
$ storm local -h ②
usage: storm local [...] jar-path main-class [...]
```

①   The packaging step creates *.asc digital signatures for all the binaries, and
       the GPG private key is used to create those signatures. **-Dgpg.skip=true**
       disables GPG signing.
②   Ensure the **storm** command is available. The **local** subcommand needs the
       path to the JAR file, main-class to run, and any arguments main-class
       will use. After uploading the JAR file, **storm local** calls the main()
       function on main-class.

4.1.1 Examine the Storm Topology Code

The *ExclamationTopology* topology code is located in the file:

```
$ cd ~/storm/examples/storm-starter/ && \
> find . -name ExclamationTopology.java
./src/jvm/org/apache/storm/starter/ExclamationTopology.java
```

The topology code imports a spout class from the file:

```
$ cd ~/storm/storm-client && find . -name TestWordSpout.java
./src/jvm/org/apache/storm/testing/TestWordSpout.java
```

The above two Java files compose a simple topology by chaining one spout and two
bolts.

Let us compile the *storm-starter* code to verify the Storm topology code for
completeness and correctness. The following command packages a JAR file ready
for submitting to a Storm cluster:

```
$ cd ~/storm/examples/storm-starter && mvn clean package
```

Maven compiles the topology Java code and packages the class files and all the
dependencies into a single "uber jar" (or "fat jar") at the path *target/storm-starter-
version.jar*. If we submit the topology JAR file without any parameter, which could
be a topology name, then the topology will be launched in local mode with a default
name "test" and run for 10 s:

```
$ cd ~/storm/examples/storm-starter/target && \
> storm local ./storm-starter-2.2.0.jar \
> org.apache.storm.starter.ExclamationTopology
```

4.1.2 Examine the Storm Flux Code

The *simple_wordcount.yaml* code is located in the *flux/flux-examples/* directory:

```
$ cd ~/storm/flux/flux-examples/ && \
> find . -name simple_wordcount.yaml
./src/main/resources/simple_wordcount.yaml
```

The YAML code depends on a spout class, which is the same spout class we saw in the previous *ExclamationTopology* topology, and two bolt classes from the files:

```
$ cd ~/storm/storm-client && find . -name TestWordSpout.java
./src/jvm/org/apache/storm/testing/TestWordSpout.java
$ cd ~/storm/storm-client && find . -name TestWordCounter.java
./src/jvm/org/apache/storm/testing/TestWordCounter.java
$ cd ~/storm/flux/flux-wrappers/src/main/java && \
> find . -name LogInfoBolt.java
./org/apache/storm/flux/wrappers/bolts/LogInfoBolt.java
```

The above three Java files and the YAML file compose a simple topology by chaining one spout and two bolts.

Let us compile the *simple_wordcount.yaml* code to verify the Storm topology code for completeness and correctness. The following command packages a JAR file ready for submitting to a Storm cluster:

```
$ cd ~/storm/flux/flux-examples && mvn clean package
```

Maven compiles the topology Java code as well as the YAML code and packages the class files, the YAML resources, and all the dependencies into a single JAR file at the path *target/flux-examples-version.jar*. The following command launches Flux topology with the *simple_wordcount.yaml* code in local mode:

```
$ storm local ./target/flux-examples-2.2.0.jar \
> org.apache.storm.flux.Flux \
> --resource /simple_wordcount.yaml
```

4.2 Migrate the Storm Topology Code to a Heron Topology Project

There are two ways to migrate a Storm topology project to a Heron topology project. One is to modify *pom.xml* in the Storm topology project, and the other is to create a new project and copy the Storm topology Java code to the new project. We use the second approach and keep the Storm project as a reference.

Create a new Maven project:

```
$ cd ~ && mvn archetype:generate
```

This command generates a Maven project directory structure interactively. It prompts for some inputs. Use the following inputs and accept the default values for the others:

```
Define value for property 'groupId': heronbook
Define value for property 'artifactId': heron-starter
```

The generated Maven project directory looks like:

```
$ tree --charset=ascii ~/heron-starter/
/home/ubuntu/heron-starter/
|-- pom.xml
`-- src
    |-- main
    |   `-- java
    |       `-- heronbook
    |           `-- App.java
    `-- test
        `-- java
            `-- heronbook
                `-- AppTest.java
```

Copy the two topology Java files that we identified in Sect. 4.1.1 to the new Maven project directory and remove the stub files and test directory since we do not have a test in this simple experiment:

```
$ cp ~/storm/examples/storm-starter/\
> src/jvm/org/apache/storm/starter/ExclamationTopology.java \
> ~/heron-starter/src/main/java/heronbook/
$ cp ~/storm/storm-client/\
> src/jvm/org/apache/storm/testing/TestWordSpout.java \
> ~/heron-starter/src/main/java/heronbook/

$ rm ~/heron-starter/src/main/java/heronbook/App.java
$ rm -rf ~/heron-starter/src/test/
```

The project directory after the adjustment looks like:

```
$ tree --charset=ascii ~/heron-starter/
/home/ubuntu/heron-starter/
|-- pom.xml
`-- src
    `-- main
        `-- java
            `-- heronbook
                |-- ExclamationTopology.java
                `-- TestWordSpout.java
```

4.2.1 Adjust the Topology Java Code

In most scenarios, the topology Java code does not need modification. In our experiment, we need to adjust the Java code in a single aspect—the package name. Since we created a new project with a new package name, we have to match the package name in the new project directory structure to the old Java code. In both Java files (*ExclamationTopology.java* and *TestWordSpout.java*), update the package name at the file header to

```
package heronbook;
```

In addition to the package name update, we have to remove the following line in *ExclamationTopology.java*, since both files are in the same package now:

~~import org.apache.storm.testing.TestWordSpout;~~

When we say that the Heron API is compatible with the Storm API, we mean the most often used stable APIs. The Heron API does not cover all the existing Storm APIs; in fact, the Heron API is a subset of the Storm API or the core set API. However, Heron's vision is fully compatible with the Storm API to make the migration easy. Therefore, the topology code may need minor adjustments to run with the Heron API, which is not necessary for this chapter.

4.2.2 Adjust the Project File pom.xml

The *pom.xml* file must include the project dependencies for a successful compilation. It also has to indicate Maven to include dependencies when packaging the JAR file.

4.2.2.1 Add Dependency

Heron provides three packages in Maven Central[2] when you search on the group ID keyword *heron*: *heron-api*, *heron-storm*, and *heron-spi* for artifact ID, respectively.

heron-api
    The native Heron API that can be used directly to build topologies.
heron-storm
    A wrapper on the native Heron API, compatible with the Storm API and depending on *heron-api*. Storm has three API series: v0.x, v1.x, and v2.x. Heron tracks the latest version of each series. We will use v2.2.0 in this chapter.
heron-spi
    The Heron Service Provider Interface used by Heron developers to extend Heron functionalities.

---

[2]https://search.maven.org/search?q=heron.

In Sect. 3.5, we saw how to compile the Heron source code. The above three API packages are generated similarly:

```
$ cd ~/heron/

$ bazel build --config=ubuntu_nostyle \
> heron/api/src/java:heron-api
$ mvn install:install-file \
> -Dfile=bazel-bin/heron/api/src/java/heron-api.jar \
> -DgroupId=heronbook -DartifactId=heron-api \
> -Dversion=0.20.3-incubating-rc7 -Dpackaging=jar

$ bazel build --config=ubuntu_nostyle \
> storm-compatibility/v2.2.0/src/java:heron-storm
$ mvn install:install-file -Dfile=\
> bazel-bin/storm-compatibility/v2.2.0/src/java/heron-storm.jar \
> -DgroupId=heronbook -DartifactId=heron-storm \
> -Dversion=0.20.3-incubating-rc7 -Dpackaging=jar

$ bazel build --config=ubuntu_nostyle \
> heron/spi/src/java:heron-spi-jar
$ mvn install:install-file \
> -Dfile=bazel-bin/heron/spi/src/java/heron-spi.jar \
> -DgroupId=heronbook -DartifactId=heron-spi \
> -Dversion=0.20.3-incubating-rc7 -Dpackaging=jar
```

These files have been installed in the local Maven repository:

```
$ ls ~/.m2/repository/heronbook
heron-api   heron-spi   heron-storm
```

We need the compatible Storm API dependency from Heron. Besides, *TestWord-Spout.java* imports the org.slf4j package, which requires a dependency item in *pom.xml*. Add the heron-storm and slf4j dependencies to *pom.xml* in the dependencies block:

```
<project>
  [...]
  <dependencies>
    <dependency>
      <groupId>heronbook</groupId>
      <artifactId>heron-storm</artifactId>
      <version>0.20.3-incubating-rc7</version>
    </dependency>
    <dependency>
      <groupId>org.slf4j</groupId>
      <artifactId>slf4j-api</artifactId>
      <version>1.7.26</version> ①
    </dependency>
    <dependency>
      <groupId>org.slf4j</groupId>
      <artifactId>slf4j-jdk14</artifactId>
      <version>1.7.26</version>
    </dependency>
    <dependency>
```

```
      <groupId>com.esotericsoftware</groupId>
      <artifactId>kryo</artifactId> ②
      <version>3.0.3</version>
    </dependency>
    [...]
</project>
```

①   The same version in the Storm root *pom.xml*.
②   Include the serialization library in the topology JAR.

4.2.2.2 Build with Dependencies

The Heron API dependencies are required to be packaged in the final topology JAR file. A few more details on why the Heron API dependencies should be embedded in the topology JAR file: when the topology JAR file is submitted through the **heron** command, the client tries to run the topology JAR file to obtain the *topology.defn* file, which is implemented by the Heron API dependencies.

The Apache Maven Assembly Plugin[3] packs the project output and its dependencies as well as other resources into a file. The plugin provides the `single` goal to create all assemblies. Add the Apache Maven Assembly Plugin in the build block to package the dependencies to the final JAR file:

```
<project>
  [...]
  <build>
    <plugins>
      <plugin>
        <artifactId>maven-assembly-plugin</artifactId> ①
        <version>3.1.0</version>
        <configuration>
          <descriptorRefs> ②
            <descriptorRef>jar-with-dependencies</descriptorRef>
          </descriptorRefs>
        </configuration>
        <executions>
          <execution>
            <phase>package</phase> ③
            <goals>
              <goal>single</goal>
            </goals>
          </execution>
        </executions>
      </plugin>
    </plugins>
    [...]
</project>
```

---

[3]http://maven.apache.org/plugins/maven-assembly-plugin/.

①   We do not need a `groupId` specification because the group is `org.apache.`
     `maven.plugins`, which is assumed by default.

②   The built-in descriptor `jar-with-dependencies` indicates a JAR of both
     the binary output and the unpacked dependencies. The generated file in our
     example will be named as *<project>-jar-with-dependencies.jar*.

③   Bind to the packaging phase.

4.2.3   Compile the Topology JAR File

Use the same command as we did for the Storm topology to package the source
code to the JAR file:

```
$ cd ~/heron-starter/ && mvn clean package
```

The generated JAR is available in the *target/* directory: *heron-starter-1.0-
SNAPSHOT-jar-with-dependencies.jar*, which is ready to be submitted to Heron
to run.

Let us take a close look at this JAR file. What is inside it:

```
$ unzip -qq \
> target/heron-starter-1.0-SNAPSHOT-jar-with-dependencies.jar \
> -d tmp && \
> tree -L 1 --charset=ascii -- tmp/
tmp/
|-- META-INF ①
|-- build-data.properties
|-- com
|-- heronbook ②
|-- javax
|-- module-info.class
`-- org
```

①   JAR metafile.

②   The topology *.class* files for *heronbook* package.

Except for the topology *.class* file and the JAR metafile, all directories and files are
from the *heron-storm* dependencies.

Try to submit this JAR file to run for a minute and then kill it:

```
$ ~/.heron/bin/heron submit local \
> target/heron-starter-1.0-SNAPSHOT-jar-with-dependencies.jar \
> heronbook.ExclamationTopology \ ①
> storm-migration-java && \ ②
> sleep 60 && pstree -aApT `ps -ef | \
> grep SchedulerMain | grep java | awk '{print $2}'` && \ ③
> heron kill local storm-migration-java
```

①   The class containing the entry point `main()`.

②   The topology name.

③   Display the topology process tree.

NOTE

The Heron CLI requires JAVA_HOME. You can set the path by

```
$ export JAVA_HOME=/usr/lib/jvm/java-11-openjdk-amd64
```

4.3   Migrate Storm Flux to Heron ECO

The Extensible Component Orchestrator (ECO) was created to build topologies without recompilation. Topology writers prepare the spouts and bolts following common procedures and obtain a JAR file. Based on the prepared spouts and bolts, the ECO uses the ECO configuration file to assemble the spouts and bolts to a DAG. The configuration file is in the YAML format and is constructed of three sections: the configuration, topology, and stream definitions [1]:

Configuration definition
> Configurations for the topology, component, and container, analogous to org. apache.heron.api.Config in the Heron low-level API.

Topology definition
> This is where you list your components, spouts, and bolts. Components will typically be configuration classes that spouts and bolts depend on. This way, you can have a spout or bolt that requires property injection other than primitives through constructor or setter injection.

Stream definition
> Defines the direction data flow between two components and the grouping type.

The ECO is an extension of Flux. Most Storm Flux topologies should be deployable in Heron with minimal changes. The ECO API is currently available to work with spouts and bolts from the package org.apache.storm, which makes migration easy.

Follow the similar procedures in Sect. 4.2, create a new Maven project, copy the Java files identified in Sect. 4.1.2 to the new project directory, adjust the Java code, and adjust the *pom.xml* file. The new project directory looks like:

```
$ cd ~ && mvn archetype:generate
...
Define value for property 'groupId': heronbook
Define value for property 'artifactId': heron-eco
...

$ cp ~/storm/storm-client/src/jvm\
> /org/apache/storm/testing/TestWordSpout.java \
> ~/heron-eco/src/main/java/heronbook/
$ cp ~/storm/storm-client/src/jvm\
> /org/apache/storm/testing/TestWordCounter.java \
```

```
>  ~/heron-eco/src/main/java/heronbook/
$ cp ~/storm/flux/flux-wrappers/src/main/java\
> /org/apache/storm/flux/wrappers/bolts/LogInfoBolt.java \
> ~/heron-eco/src/main/java/heronbook/

$ mkdir -p ~/heron-eco/src/main/resources
$ cp ~/storm/flux/flux-examples/src/main/resources\
> /simple_wordcount.yaml ~/heron-eco/src/main/resources/

$ rm ~/heron-eco/src/main/java/heronbook/App.java
$ rm -rf ~/heron-eco/src/test/

$ cp ~/heron-starter/pom.xml ~/heron-eco/pom.xml
$ sed -i 's/heron-starter/heron-eco/g' ~/heron-eco/pom.xml

$ tree --charset=ascii ~/heron-eco/
/home/ubuntu/heron-eco/
|-- pom.xml
`-- src
    `-- main
        |-- java
        |   `-- heronbook
        |       |-- LogInfoBolt.java
        |       |-- TestWordCounter.java
        |       `-- TestWordSpout.java
        `-- resources
            `-- simple_wordcount.yaml
```

Besides the package name update in the Java files, we have to update the package
name for the spout and bolts from org.apache.storm._ to heronbook._ in the
YAML file accordingly:

```
className: "heronbook.TestWordSpout"
className: "heronbook.TestWordCounter"
className: "heronbook.LogInfoBolt"
```

Build and find the JAR file under *target/* and examine the JAR file to make sure
that the ECO class is included:

```
$ cd ~/heron-eco/ && mvn clean package
$ unzip -qq -d tmp \
> target/heron-eco-1.0-SNAPSHOT-jar-with-dependencies.jar
$ ls tmp/*yaml
tmp/simple_wordcount.yaml  (1)
$ find tmp -name Eco.class
tmp/org/apache/heron/eco/Eco.class  (2)

$ heron submit local \
> target/heron-eco-1.0-SNAPSHOT-jar-with-dependencies.jar \
> org.apache.heron.eco.Eco \
> --eco-config-file tmp/simple_wordcount.yaml && \  (3)
> sleep 60 && pstree -aApT `ps -ef | \
> grep SchedulerMain | grep java | awk '{print $2}'` && \
> heron kill local yaml-topology
```

Table 4.1 Storm and Heron compiling comparison

Features	Storm	Heron
API	Storm API	core Storm API
pom.dependency	storm	heron-storm
pom.jar-with-dependency	N	Y

① The JAR file includes the Heron ECO configuration file.
② The class to be launched with.
③ Indicate the ECO YAML file location.

4.4 Summary

During the previous migration process, we found that the Storm topology and Heron topology are different in several aspects: the API, the *pom.xml*, and the output JAR file, summarized in Table 4.1.

In this chapter, we started from the *storm-starter* project and showed how Storm compiles the topology. Then, we created a Heron project, copied the Storm topology Java code to the new Heron project, and adjusted the Java code. We edited the project file *pom.xml* to add dependency and plugins for compiling options. Finally, we generated a topology JAR file ready to be submitted to Heron. Following a similar process, we migrated a Storm Flux topology to the Heron ECO topology as well.

References

1. Introducing Heron's ECO; a flexible way to manage topologies. https://1904labs.com/2018/02/14/introducing-herons-eco-flexible-way-manage-topologies/. Visited on 2020-07
2. Flux. https://storm.apache.org/releases/current/flux.html. Visited on 2020-07

Chapter 5
Write Topology Code

In Chap. 4, we migrated the existing Storm topologies to Heron successfully without writing the topology codes ourselves. However, you can write your new topologies. This chapter will show you how to compose a new topology with Heron low-level Topology APIs in Java and Python.

Besides topology composition, you may be curious about the answers to these questions: How do topologies run? How are tuples transmitted from one vertex to another vertex in the directed acyclic graph (DAG)? What does the whole tuple data flow look like? We will dive deep into a topology to find the answers by experiments from the topology writer's perspective.

5.1 Before Writing Code

Before working on code, let us first design a topology that includes the most often seen topology patterns in Sect. 5.1.1. We also have to choose a programming language to code the designed topology. Heron provides three basic choices: Java, Python, and C++. We will discuss the Heron API hierarchy in Sect. 5.1.2.

5.1.1 Design Topology

In Sect. 4.1.1, we saw the *ExclamationTopology* topology. We revisit this well-known typical topology, and then we will design our own new topology. In its `main()`, the topology was constructed by

```
builder.setSpout("word", new TestWordSpout(), 10);
builder.setBolt("exclaim1", new ExclamationBolt(), 3)
        .shuffleGrouping("word");
builder.setBolt("exclaim2", new ExclamationBolt(), 2)
        .shuffleGrouping("exclaim1");
```

© The Author(s), under exclusive license to Springer Nature Switzerland AG 2021
H. Wu, M. Fu, *Heron Streaming*, https://doi.org/10.1007/978-3-030-60094-5_5

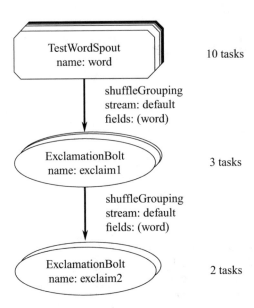

Fig. 5.1 *ExclamationTopology* directed acyclic graph

The topology builder sets one spout and two bolts. They are linked as three tiers by shuffle grouping, as shown in Fig. 5.1. All the streams are named *default*, which is assigned by Heron if the topology writer does not specify a name. The stream schema is one field named *word*, which is specified in the `TestWordSpout` and `ExclamationBolt` classes.

The *ExclamationTopology* topology demonstrates how to link multiple nodes in a graph. However, there is only one stream path, which cannot show how a node handles multiple input or output streams. We can add one more stream path to the existing topology, which looks like Fig. 5.2. Now, there are two streams between spout *word* and bolt *exclaim1*. Both the streams are shuffle grouping: one of them is named *s1* and the other *s2*. Stream *s1* has one field, *f1*, while stream *s2* has two fields, *f2* and *f3*. In this topology, there are two tuple paths:

- *word* → *s1*:[*f1*] → *exclaim1* → *default*:[*word*] → *exclaim2*
- *word* → *s2*:[*f2,f3*] → *exclaim1* → *default*:[*word*] → *exclaim2*

From a topology writer's perspective, compared to the original topology DAG, we expect code modifications so that the spout can emit the second output stream and the bolt can handle the second input stream. Besides the code change on the topology structure, we add logs to trace the tuple flow. Since all the processes are on the same machine in our experiment, topology task processes may directly write to the same log file, and we can observe the topology running footprint by applying `tail -f` on the log file.

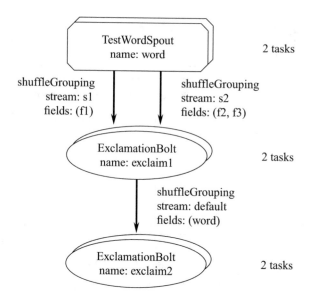

Fig. 5.2 Modified *ExclamationTopology* DAG

5.1.2   Choose a Heron API

Core Heron APIs depend on Heron Instance implementations. There are at present three Heron Instance implementations of Java, Python, and C++. Corresponding to the Heron Instance implementations are three low-level Topology APIs shown in Fig. 5.3. Since there is no language translation or conversion for the low-level API languages, the performance for each language is always best when run on its own. The topology developers may choose the most familiar language among the three languages to compose the topology.

The Heron APIs can be categorized into two sets: low-level APIs and high-level APIs. The categorization criterion is the programming data model. The low-level or Topology API can describe a directed acyclic graph with spout and bolt as vertexes and streams as edges. The high-level or streamlet or functional API hides the topology details but can describe stream manipulation such as map, join, and consume. We will see the high-level API data model in Chap. 7.

Since the Heron Topology API data model is almost the same as or compatible with Storm, it is easy to wrap a thin layer to convert the Storm Java API invocation to the Heron Java API invocation. The Heron ECO may be considered as a YAML API that extends Heron and Storm Java APIs. In the following two sections, we will use Java and Python low-level Topology APIs to code the proposed topology design in Sect. 5.1.1.

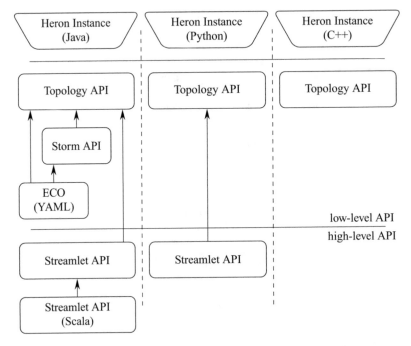

Fig. 5.3 Heron API overview

5.2 Write Topology in Java

In this section, we modify the topology code in Sect. 4.2 to fulfill the design in
Sect. 5.1.1. Copy the new project:

```
$ cp -r ~/heron-starter ~/heron-topology-java
$ cd ~/heron-topology-java && mvn clean && rm -rf tmp
$ sed -i 's/heron-starter/heron-topology-java/g' pom.xml
```

From now on, we will focus on Heron APIs. It is easy to find the corresponding
Heron classes from Storm classes by searching them in the Heron Javadoc.[1] Change
all the import org.apache.storm._ to import org.apache.heron.api._ in
ExclamationTopology.java:

```
import org.apache.heron.api.Config;
import org.apache.heron.api.HeronSubmitter;
import org.apache.heron.api.bolt.BaseRichBolt;
import org.apache.heron.api.bolt.OutputCollector;
import org.apache.heron.api.topology.OutputFieldsDeclarer;
import org.apache.heron.api.topology.TopologyBuilder;
import org.apache.heron.api.topology.TopologyContext;
```

---

[1] https://heron.incubator.apache.org/api/java/.

```
import org.apache.heron.api.tuple.Fields;
import org.apache.heron.api.tuple.Tuple;
import org.apache.heron.api.tuple.Values;
```

Similarly, update the imports in *TestWordSpout.java*:

```
import org.apache.heron.api.spout.BaseRichSpout;
import org.apache.heron.api.spout.SpoutOutputCollector;
import org.apache.heron.api.topology.OutputFieldsDeclarer;
import org.apache.heron.api.topology.TopologyContext;
import org.apache.heron.api.tuple.Fields;
import org.apache.heron.api.tuple.Values;
import org.apache.heron.api.utils.Utils;
```

5.2.1   Code the Topology

To write logs, we need a logging library. Heron itself uses *slf4j* in Java code. We could use *slf4j* as well in our experiment. However, we do not want to mix our logs with Heron logs; thus, we pick a different simple log library called *tinylog*[2] in the experiment. *tinylog* claims it can handle multiple processes writing to a shared log file, which is good for us to collate logs in one place when we run the topology in local mode.

Besides the log library, we also use the library *console-table-builder*[3] to print a nice table in text format for easy reading. Add the log and table library dependencies in the project file *pom.xml*:

```
<dependencies>
    ...
    <dependency>
      <groupId>heronbook</groupId>
      <artifactId>heron-api</artifactId>  (1)
      <version>0.20.3-incubating-rc7</version>
    </dependency>
    <dependency>
      <groupId>org.tinylog</groupId>
      <artifactId>tinylog</artifactId>  (2)
      <version>1.3.6</version>
    </dependency>
    <dependency>
      <groupId>io.bretty</groupId>
      <artifactId>console-table-builder</artifactId>  (3)
      <version>1.2</version>
    </dependency>
    ...
</dependencies>
```

---

[2]https://tinylog.org/.

[3]https://mvnrepository.com/artifact/io.bretty/console-table-builder.

① Although you can keep *heron-storm* as it includes the Heron low-level API, we change to *heron-api* for a clean dependency.

② A log library, which is different from *slf4j*.

③ A table formatting library to print well-formatted text tables.

5.2.1.1 Code Main

Since the helper, `ConfigurableTopology`, is not available in the Heron API, we remove its inheritance and merge `run()` into `main()`. As for the topology structure, since most parts are the same as the original topology, we keep most of the code unchanged when we wire the spout and bolts together. The only change is one more stream between spout "word" and bolt "exclaim1". We append below `shuffleGrouping()` to "exclaim1" with stream names in `main()` of *ExclamationTopology.java*:

```
public static void main(String[] args) throws Exception {
  TopologyBuilder builder = new TopologyBuilder();
  builder.setSpout("word", new TestWordSpout(), 2);
  builder.setBolt("exclaim1", new ExclamationBolt(), 2)
         .shuffleGrouping("word", "s1")
         .shuffleGrouping("word", "s2"); ①
  builder.setBolt("exclaim2", new ExclamationBolt(), 2)
         .shuffleGrouping("exclaim1");

  Config conf = new Config();
  conf.setDebug(true);
  conf.setNumStmgrs(3);

  String topologyName = "test";
  if (args != null && args.length > 0) {
    topologyName = args[0];
  }

  HeronSubmitter.submitTopology(
    topologyName, conf, builder.createTopology()); ②
}
```

① The bolt `exclaim1` subscribes two shuffle grouping streams, `s1` and `s2`, from the spout `word`.

② Submit the topology with the name `topologyName`.

5.2.1.2 Code Spout

When the topology JAR file is submitted through the Heron CLI, the CLI tries to parse the topology structure. The key information to construct the topology graph for the spout is the output stream schema, which is defined in `declareOutputFields()` of *TestWordSpout.java*:

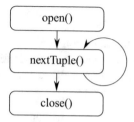

Fig. 5.4 Spout life cycle

```
public void declareOutputFields(OutputFieldsDeclarer declarer) {
    declarer.declareStream("s1", new Fields("f1"));
    declarer.declareStream("s2", new Fields("f2", "f3"));
}
```

When a container is running, and the spout task is loaded, the spout life cycle starts. The spout life cycle consists of three stages: open(), loop of nextTuple(), and close(), as shown in Fig. 5.4.

At the open() stage, we need to initialize the logging, followed by optionally logging the stream schema defined in declareOutputFields():

```
public void open(Map conf, TopologyContext context,
                 SpoutOutputCollector collector) {
    this.collector = collector;
    setLoggging(context);
    Logger.trace("streams: {}", context.getThisStreams());
}
```

Logger was picked from the *tinylog* library. Thus, we remove the unused slf4j.Logger imports. Instead, remember to import *tinylog* classes.

```
import org.pmw.tinylog.*;
import org.pmw.tinylog.writers.*;

import org.slf4j.Logger;
import org.slf4j.LoggerFactory;

public static Logger LOG = ...
```

The logging setup details are as follows:

```
public static void setLoggging(TopologyContext context) {
    LoggingContext.put("tId", context.getThisTaskId());      ①
    LoggingContext.put("cId", context.getThisComponentId()); ②
    LoggingContext.put("tIdx", context.getThisTaskIndex());  ③
    Configurator.currentConfig()
        .formatPattern("{date:yyyy-MM-dd HH:mm:ss} "
            + "/{context:tId}({context:cId}[{context:tIdx}])/ " ④
            + "[{thread}]\n{class}.{method}() {level}:\n{message}")
        .writer(new SharedFileWriter("/tmp/log.txt", true),  ⑤
                Level.TRACE)
        .activate();
}
```

①   The task ID among all the nodes in the topology, or the global task ID.
②   The component name.
③   The task index under the same component name.
④   The log leading string indicates the task identity. A task or Heron Instance can be identified by either `tId` or the `<cId, tIdx>` pair.
⑤   The log is located at */tmp/log.txt*.

At the `nextTuple()` stage, we have to generate tuples for the two streams. Besides, we do not want the tuple generation speed to be too fast to make the log hard to read. `TestWordSpout.java:nextTuple()` looks like:

```
public void nextTuple() {
  Utils.sleep(1000);  ①
  final String[] words = new String[] {
    "nathan", "mike", "jackson", "golda", "bertels"};
  final Random rand = new Random();
  final String w1 = words[rand.nextInt(words.length)];  ②
  final String w2 = words[rand.nextInt(words.length)];
  final String w3 = words[rand.nextInt(words.length)];
  final Values v1 = new Values(w1);  ③
  final Values v2 = new Values(w2, w3);
  final Integer msgId = Integer.valueOf(rand.nextInt());  ④
  collector.emit("s1", v1, msgId);  ⑤
  collector.emit("s2", v2);
  Logger.trace("emit {} {} {}", "s1", v1, msgId);  ⑥
  Logger.trace("emit {} {}", "s2", v2);
}
```

①   Make the tuple generation speed slow, so that we can read the logs when the topology is running.
②   Step 1: Prepare three words.
③   Step 2: Prepare two values, one for each stream.
④   Message ID is required to enable acking or at-least-once semantics. In the experiment, we enable acking in stream *s1*, while keeping stream *s2* in at-most-once semantics.
⑤   Step 3: Emit to two streams.
⑥   Step 4: Log the tuples.

At the `close()` stage, as well as the other methods `ack()` and `fail()`, we add logging, so that we get notified when they are called. Moreover, we remove `getComponentConfiguration()` because its implementation is not compatible with the Heron API.

5.2.1.3   Code Bolt

We follow similar steps to modify the bolt code. Since the output schema does not change, `declareOutputFields()` remains unchanged. Similarly to the spout, the

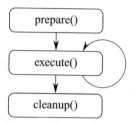

Fig. 5.5 Bolt life cycle

bolt life cycle consists of `prepare()`, a loop of `execute()`, and `cleanup()`, as shown in Fig. 5.5.

At the `prepare()` stage, we initialize logging and log the stream name, which is similar to what we did in the spout `open()` stage. Remember to import `tinylog` classes.

```
import org.pmw.tinylog.*;

public void prepare(Map conf,
                    TopologyContext context,
                    OutputCollector collector) {
  this.collector = collector;
  TestWordSpout.setLoggging(context);
  Logger.trace("streams: {}", context.getThisStreams());
}
```

At the `execute()` stage, we have to handle the two input streams:

```
public void execute(Tuple tuple) {
  String srcComponent = tuple.getSourceComponent();
  String srcStream    = tuple.getSourceStreamId();
  int    srcTask      = tuple.getSourceTask();
  Logger.trace("received tuple from `{}` of /{}({})/\n{}",
    srcStream, srcTask, srcComponent, tupleAsTable(tuple)); ①

  if (srcStream .equals("s2")) { ②
    final String v = tuple.getStringByField("f2") + "&" +
                     tuple.getStringByField("f3") + "!!!"; ③
    collector.emit(new Values(v)); ④
  } else { ⑤
    collector.emit(tuple,
                   new Values(tuple.getString(0) + "!!!"));
    collector.ack(tuple); ⑥
  }
}
```

①   Print the input tuple data, including the source component, the source stream, and the source task ID. The <srcComponent, srcStream> pair identifies a stream.

②   A separate branch handles stream s2.

(3)   Prepare the value of the tuple to be emitted: combine the two values in the two fields into one.

(4)   Emit the tuple for stream s2.

(5)   The branch for stream s1, which is the same as the original code of execute().

(6)   The acking (more details in Sect. 6.1.1) is done for the tuples from the stream path including s1, since we only enable acking for the tuple path *word* → *s1*:[*f1*] → *exclaim1* → *default*:[*word*] → *exclaim2* in our previous design in Sect. 5.1.1. When the bolt exclaim2 receives a tuple on the stream path including s2, it should not ack that tuple. However, exclaim1 and exclaim2 share the code, and all tuples are acked in this branch in exclaim2. Fortunately, due to the acking feature, if an unanchored tuple is used as an anchor, anchoring is ignored. Thus, it is fine to ack the tuple on the stream path including s2.

tupleAsTable() prints the tuple as a two-row table for easy reading, thanks to the *console-table-builder* library. Remember to import *console-table-builder* classes as well as HashMap.

```java
import java.util.HashMap;
import io.bretty.console.table.*;

private static String tupleAsTable(Tuple tuple) {
  Map<Integer, String> tableHeader =
    new HashMap<Integer, String>();
  for (String s: tuple.getFields()) {
    tableHeader.put(tuple.fieldIndex(s), s);
  } // table head
  Object[][] data = new Object[2][tuple.size()];
  for (int i=0; i<tuple.size(); i++) {
    data[0][i] = tableHeader.get(i);
    data[1][i] = tuple.getValue(i);
  } // table body
  return Table.of(data, Alignment.LEFT, 20).toString();
}
```

5.2.2   Understand Tuple Flow

Up until now, we focused on updating the code. Next, we follow the same steps in Sect. 4.2.3 to compile the code and the same steps in Sect. 2.4 to run the topology. After the topology is launched, we expect that the **tail -f** command will show the tuple and stream logs for us to analyze.

```
$ cd ~/heron-topology-java/ && mvn clean package

$ heron submit local ./target/heron-topology-java\
> -1.0-SNAPSHOT-jar-with-dependencies.jar \
> heronbook.ExclamationTopology my-java-topology && \
> sleep 60 && tail -f /tmp/log.txt
```

The `my-java-topology` topology is our first, basic, simple topology, which will be referenced multiple times in later chapters. To stop this running topology, use the command:

```
$ heron kill local my-java-topology
```

5.2.2.1 How Tuple Is Constructed

Here is an excerpt from */tmp/log.txt*, which shows four tuples:

```
2020-07-19 13:49:08 /3(exclaim1[0])/ [SlaveThread]
heronbook.ExclamationTopology$ExclamationBolt.execute() TRACE:
received tuple from `s1` of /5(word)/
|f1            |
|jackson       |

2020-07-19 13:49:09 /3(exclaim1[0])/ [SlaveThread]
heronbook.ExclamationTopology$ExclamationBolt.execute() TRACE:
received tuple from `s2` of /5(word)/
|f2            |f3          |
|jackson       |nathan      |

2020-07-19 13:49:09 /1(exclaim2[0])/ [SlaveThread]
heronbook.ExclamationTopology$ExclamationBolt.execute() TRACE:
received tuple from `default` of /3(exclaim1)/
|word             |
|jackson&nathan!!!|

2020-07-19 13:49:09 /2(exclaim2[1])/ [SlaveThread]
heronbook.ExclamationTopology$ExclamationBolt.execute() TRACE:
received tuple from `default` of /3(exclaim1)/
|word          |
|jackson!!!    |
```

We found that the tuple is constructed as the field-value or key-value. The field/key index is the same as defined in the method `declareOutputFields()`. The value matches the count and order of the field/key.

Analogous to the table concept, each stream is like an endless table. The table head is defined as fields/keys, and each row in the table is a tuple injected to the stream each time `emit()` is called. The `emit()` method is like a cursor scanning the table, as shown in Fig. 5.6.

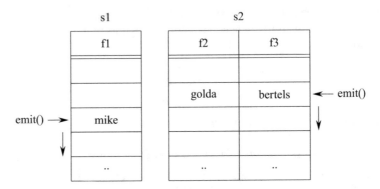

Fig. 5.6 Analogy to table scanning

5.2.2.2 How Tuple Is Routed

The two tuple paths in the topology can be confirmed in the log. The following log excerpts show the tuple paths.

The first case shows the path *word* → *s1:[f1]* → *exclaim1* → *default:[word]* → *exclaim2*:

```
2020-07-19 13:49:08 /5(word[0])/ [SlaveThread]  ①
heronbook.TestWordSpout.nextTuple() TRACE:
emit s1 [jackson] -355585852

2020-07-19 13:49:08 /3(exclaim1[0])/ [SlaveThread]  ②
heronbook.ExclamationTopology$ExclamationBolt.execute() TRACE:
received tuple from `s1` of /5(word)/
|f1                        |
|jackson                   |

2020-07-19 13:49:09 /2(exclaim2[1])/ [SlaveThread]  ③
heronbook.ExclamationTopology$ExclamationBolt.execute() TRACE:
received tuple from `default` of /3(exclaim1)/
|word                      |
|jackson!!!                |
```

①   The spout 5(word[0]) emitted tuple jackson with a message ID -355585852 into stream s1.

②   The bolt 3(exclaim1[0]) received tuple jackson from stream s1 of the Heron Instance with global task ID 5, then it appended !!!.

③   The bolt 2(exclaim2[1]) received tuple jackson!!! from the stream default of the Heron Instance with global task ID 3.

The second case shows the path *word* → *s2:[f2,f3]* → *exclaim1* → *default:[word]* → *exclaim2*:

```
2020-07-19 13:49:08 /5(word[0])/ [SlaveThread] ①
heronbook.TestWordSpout.nextTuple() TRACE:
emit s2 [jackson, nathan]

2020-07-19 13:49:09 /3(exclaim1[0])/ [SlaveThread] ②
heronbook.ExclamationTopology$ExclamationBolt.execute() TRACE:
received tuple from `s2` of /5(word)/
|f2                |f3                |                 |
|jackson           |nathan            |                 |

2020-07-19 13:49:09 /1(exclaim2[0])/ [SlaveThread] ③
heronbook.ExclamationTopology$ExclamationBolt.execute() TRACE:
received tuple from `default` of /3(exclaim1)/
|word              |                 |
|jackson&nathan!!! |                 |
```

①   The spout `5(word[0])` emitted tuple `[jackson, nathan]` to stream `s2`. There is no message ID, which means the spout did not expect acking for this tuple.

②   The bolt `3(exclaim1[0])` received the tuple from stream `s2` of the Heron Instance with global task ID 5.

③   The bolt `1(exclaim2[0])` received the tuple from the stream `default` of the Heron Instance with global task ID 3.

5.3 Write Topology in Python

Besides Java, topology writers can also use Python 3. We use Pants[4] to manage the Python topology code. Pants makes the manipulation and distribution of hermetically sealed Python environments painless. You can organize your code in the Pants way with targets for binaries, libraries, and tests. Pants builds Python code into PEXes. A PEX[5] is, roughly, an archive file containing a runnable Python environment [1]. Set up the development environment first:

```
$ mkdir -p ~/heron-topology-py && cd ~/heron-topology-py ①
$ curl -L -O https://pantsbuild.github.io/setup/pants && \
> chmod +x pants ②
```

①   Make the project directory.

②   Download the **pants** code management tool and make it executable.

---

[4]https://www.pantsbuild.org/docs.

[5]https://legacy.python.org/dev/peps/pep-0441/.

In this simple example, we put all files in the source root directory. Eventually, we will have the following files before compiling:

```
$ tree --charset=ascii ~/heron-topology-py/
/home/ubuntu/heron-topology-py/
|-- BUILD
|-- bolt.py
|-- heronpy-0.0.0-py3-none-any.whl
|-- pants
|-- pants.toml
|-- spout.py
`-- topology.py
```

Edit the *pants.toml* file to indicate the **pants** version and the source root directory as the local dependency repository:

```
[GLOBAL]
pants_version = "1.29.0"
[python-repos]
repos = ["%(buildroot)s/"]
```

We can build a Python API package from the local source code using the following commands, which we already saw in Sect. 3.5. The Python API package is in the *wheel* file, which we copy to the project directory to match the configured local Python dependency repository in *pants.toml*.

```
$ cd ~/heron/ && bazel build \
> --config=ubuntu_nostyle scripts/packages:pypkgs  ①
$ cp bazel-bin/scripts/packages/heronpy.whl \
> ~/heron-topology-py/heronpy-0.0.0-py3-none-any.whl  ②
```

①   Build the Python API package, the same as in Sect. 3.5.
②   Copy the generated wheel file to the project directory.

Pants needs the project file to describe the source code files and dependencies, which is similar to Bazel. The project file *BUILD* looks like:

```
python_requirement_library(  ①
  name = 'python_requirement_library',
  requirements = [python_requirement('heronpy==0.0.0')]  ②
)

python_library(  ③
  name = 'spout-bolt-py',
  sources = ['*.py', '!topology.py']
)

python_binary(  ④
  name = "exclamation-topology",
  sources = ['topology.py'],  ⑤
  dependencies = [':spout-bolt-py',
                  ':python_requirement_library']  ⑥
)
```

①   The dependency library: depends on *heronpy*.
②   In our example, the version is 0.0.0. Alternatively, you can search the desired Python API version in the Python Package Index (PyPI) repository.[6] Then, replace 0.0.0 with the target version.
③   The library including spouts and bolts code.
④   Topology PEX file.
⑤   Topology main entry.
⑥   Link the binary to the *heronpy* library.

5.3.1   Code Main

There are two ways to compose a Heron Python topology once spouts and bolts are ready: use the TopologyBuilder class in the main function, or subclass the Topology class.

We use the designed topology in Sect. 5.1.1 as an example to write the Python topology. We use the TopologyBuilder class in the main function to create the topology, as shown in the following file *topology.py*:

```
from heronpy.api.stream import Grouping
from heronpy.api.topology import TopologyBuilder ①

from spout import TestWordSpout
from bolt import ExclamationBolt

if __name__ == '__main__':
  builder = TopologyBuilder('my-python-topology') ②
  word = builder.add_spout('word', TestWordSpout, par=2) ③
  exclaim1 = builder.add_bolt('exclaim1', \ ④
    ExclamationBolt, par=2, \
    inputs={word['stream1']: Grouping.SHUFFLE, \
            word['stream2']: Grouping.SHUFFLE}) ⑤
  exclaim2 = builder.add_bolt('exclaim2', \
    ExclamationBolt, par=2, \
    inputs={exclaim1: Grouping.SHUFFLE})
  builder.build_and_submit() ⑥
```

①   Import Grouping and TopologyBuilder from heronpy.
②   In the main function, we instantiate a TopologyBuilder with the topology name. We hard-code the topology name in this simple experiment. The TopologyBuilder object exposes two methods—add_bolt and add_spout—for adding bolts and spouts. Both methods return the corresponding HeronComponentSpec object.
③   add_spout takes three arguments and an optional dict configuration parameter. The first argument assigns a unique identifier for this spout. The

---

[6]https://pypi.org/project/heronpy/.

second argument indicates the `Spout` subclass of user logic code. The third argument `par` specifies the parallelism.

(4)   `add_bolt` takes four arguments and an optional `dict` configuration parameter. The first three arguments are the same with `add_spout`. The fourth argument indicates the bolt's inputs.

(5)   `inputs` could be a dictionary mapping from `HeronComponentSpec` to `Grouping` or a list of `HeronComponentSpecs` with default shuffle grouping.

(6)   Finalize the topology graph.

  `TopologyBuilder` also exposes the `set_config` method for specifying the topology-wide configuration, which is `dict`, and its keys are defined in the `api_constants` module.

5.3.2   Code Spout

To create a spout, subclass the `Spout` class, which has the following methods:

- The `initialize()` method is called when the spout is initialized and provides the spout with the executing environment. It is equivalent to the `open()` method of `ISpout`. Note that you should not override the `__init__()` constructor of the `Spout` class for the initialization of custom variables, since it is used internally by `HeronInstance`; instead, `initialize()` should be used to initialize any custom variables or connections to databases.
- The `next_tuple()` method is used to fetch tuples from the input source. You can emit fetched tuples by calling `self.emit()`, as described below.
- The `ack()` method is called when `HeronTuple` with a tup_id emitted by this spout is successfully processed.
- The `fail()` method is called when `HeronTuple` with a tup_id emitted by this spout is not processed successfully.
- The `activate()` method is called when the spout is asked to change to the active state.
- The `deactivate()` method is called when the spout is asked to enter the deactivated state.
- The `close()` method is called when the spout is shut down. There is no guarantee that this method will be called due to how the instance is killed.

The `Spout` class inherits from the `BaseSpout` class, which also provides methods you can use in your spouts:

- The `emit()` method is used to emit a given tuple, which can be a list or tuple of any Python objects. Unlike in the Java implementation, there is no `OutputCollector` in the Python implementation.
- The `log()` method is used for logging an arbitrary message, and its outputs are redirected to the log file of the component. It accepts an optional argument that specifies the logging level. By default, its logging level is INFO. Due to an

internal issue, you should NOT output anything to `sys.stdout` or `sys.stderr`; instead, you should use this method to log anything you want.

- In order to declare the output fields of this spout, you need to place class attribute outputs as a list of `str` or `Stream`. Note that unlike Java, `declareOutputFields` does not exist in the Python implementation. Moreover, you can optionally specify the output fields from the `spec()` method from the `optional_outputs`.
- You can use the `spec()` method to define a topology and specify the location of this spout within the topology, as well as set up component-specific configurations.

spout.py looks like:

```
from heronpy.api.spout.spout import Spout
from heronpy.api.stream import Stream    ①
from time import sleep
from random import choice

class TestWordSpout(Spout):    ②
  words = ('nathan', 'mike', 'jackson', 'golda', 'bertels')
  outputs = (    ③
    Stream(fields=('field1',), name='stream1'),
    Stream(fields=('field2', 'field3'), name='stream2')
  )

  def next_tuple(self):    ④
    sleep(5)    ⑤
    word1 = choice(self.words)    ⑥
    word2 = choice(self.words)
    word3 = choice(self.words)
    self.log('next_tuple stream1 ' + word1)    ⑦
    self.emit(tup=(word1,), stream="stream1")
    self.log('next_tuple stream2 ' + word2 + ' ' + word3)
    self.emit(tup=(word2, word3), stream="stream2")
```

①	Import the spout type from `heronpy`.
②	`TestWordSpout` that inherits from the Heron spout.
③	Important: Define output field tags for the spout.
④	Generate the next tuple sequence for this spout.
⑤	Sleep for 5 s to throttle spout.
⑥	Get the next random word.
⑦	Emit word to go to the next phase in the topology.

5.3.3   Code Bolt

Bolts must inherit the `Bolt` class, which is similar to `Spout` except that `next_tuple()` is replaced by `process()`. The `process()` method is called to process a single input tuple of `HeronTuple` type. This method is equivalent to the execute() method of the IBolt interface in Java. You can use the self.emit() method to emit the result, as described below.

Besides, Bolt's superclass BaseBolt is similar to BaseSpout with the following additional methods:

- The ack() method is used to indicate that the processing of a tuple has succeeded.
- The fail() method is used to indicate that the processing of a tuple has failed.
- The is_tick() method returns whether a given tuple of the HeronTuple type is a tick tuple.

bolt.py looks like:

```
from heronpy.api.bolt.bolt import Bolt  ①

class ExclamationBolt(Bolt):  ②
  outputs = ['word']  ③
  def process(self, tup):  ④
    if tup.stream == 'stream2':
      word = tup.values[0]+'&'+tup.values[1]+'!!!!'  ⑤
      self.emit((word,))  ⑥
    else:
      word = tup.values[0] + '!!!!'
      self.emit((word,))
    self.log('process ' + word)
```

①	Import Bolt type from heronpy.
②	ExclamationBolt inherits from the Heron Bolt.
③	Important: define output field tags for the bolt.
④	Process the word stream to aggregate to word count.
⑤	Prepare the word to emit.
⑥	Emit word as a tuple.

5.3.4   Compile and Run

To compile the source code to a PEX file, run the following command, which generates the PEX in the *./dist* directory. Run the following topology:

```
$ cd ~/heron-topology-py && \
> ./pants clean-all && ./pants binary //:  ①

$ heron submit local ./dist/exclamation-topology.pex - && \  ②
> sleep 120 && heron kill local my-python-topology
```

| ① | *<goal>* is binary; *<target>* is //:. To get more details on command usage, run **./pants help**. |
| ② | No class definition as this is a Python package. We do not feed the topology name, which is hard-coded in *topology.py*. |

5.4   Summary

In this chapter, we modified the example topology code by adding one more stream path. We also added logs that helped us analyze how the tuple was constructed and how the stream path worked. The job was updated to depend on the Heron API instead of the Storm API. Moreover, we wrote our first Python topology. In the next chapter, we will study more topology features.

Reference

1. Python projects with Pants. https://v1.pantsbuild.org/python_readme.html. Visited on 2020-07

Chapter 6
Heron Topology Features

In the last chapter, we saw how to write a simple topology. The Heron API supports more advanced concepts, some of which we will study in this chapter. When tuples flow through a topology, some tuples may fail. Heron provides two more advanced tuple delivery semantics besides the default naive at-most-once. We will discuss at-least-once and effectively-once in Sect. 6.1. In some application scenarios, the incoming data is bounded by windows. Heron supports various windowing operations covered in Sect. 6.2.

6.1 Delivery Semantics

Delivery semantics is set for each topology. The default value is at-most-once if you do not set anything. According to your application scenario, you can also set more advanced semantics such as at-least-once and effectively-once.

Heron supports three delivery semantics: at-most-once, at-least-once, and effectively-once. Table 6.1 [1] defines each. The reliability modes are implemented by different mechanisms. Table 6.2 summarizes the reliability mode implementation mechanisms.

6.1.1 At-Least-Once

To enable at-least-once in the topology, you have to do the following:

- Set the reliability mode to `TopologyReliabilityMode.ATLEAST_ONCE`. This enum turns on the at-least-once mode.
- Set the tuple ID in the spout `emit()` when you send it out. The tuple ID is used to track this tuple. If the tuple fails for any reason, the spout will try to replay

© The Author(s), under exclusive license to Springer Nature Switzerland AG 2021
H. Wu, M. Fu, *Heron Streaming*, https://doi.org/10.1007/978-3-030-60094-5_6

Table 6.1 Available semantics

Semantics	Description	When to use?
At-most-once	Heron processes tuples using a best-effort strategy. With at-most-once semantics, some of the tuples delivered into the system may be lost due to some combination of processing, machine, and network failures. What sets at-most-once semantics apart from the others is that Heron will not attempt to retry a processing step upon failure, which means that the tuple may fail to be delivered.	When some amount of data loss is acceptable
At-least-once	Tuples injected into a Heron topology are guaranteed to be processed at least once; no tuple will fail to be processed. It is possible, however, that any given tuple is processed more than once in the presence of various failures, retries, or other contingencies.	When you need to guarantee no data loss
Effectively-once	Heron ensures that the data it receives is processed effectively once—even in the presence of various failures—leading to accurate results. This applies only to stateful topologies. "Effectively," in this case, means that there is a guarantee that tuples that cause state changes will be processed once (i.e., they will affect the state once).	When you are using stateful topologies and need a strong no-data-loss guarantee

Table 6.2 Reliability mode implementation mechanisms

Semantics	Acking	Stateful
At-most-once	No	No
At-least-once	Yes	No
Effectively-once	No	Yes

it. There are two common reasons for tuple failure: timeout and the `fail()` invocation. If the spout fails to collect the acking message for a particular tuple within a configured time threshold, it considers that as a tuple timeout. The default timeout threshold is 30 s. You can change it in the configuration. If the downstream bolt invokes `fail()` to fail the tuple explicitly, the spout will know as well.

- Call `ack()` or `fail()` when you finish processing tuple. If you forget to ack a tuple, the corresponding spout will receive the timeout on this tuple eventually.

These three items have to be fulfilled together to turn on at-least-once delivery semantics. Take the Java topology in Sect. 5.2.1 as an example to create a demonstration project:

```
$ cp -r ~/heron-topology-java ~/heron-topology-java-at-least-once
$ sed -i 's/-topology-java/-topology-java-at-least-once/g' \
> ~/heron-topology-java-at-least-once/pom.xml
```

The following code shows how to enable at-least-once during topology construction and bolt implementation:

```
public static void main(String[] args) throws Exception {
  ...
  Config conf = new Config();
  conf.setMessageTimeoutSecs(10);  ①
  conf.setTopologyReliabilityMode(
    Config.TopologyReliabilityMode.ATLEAST_ONCE);  ②
  conf.setDebug(true);
  ...
}

public static class ExclamationBolt extends BaseRichBolt {
  public void execute(Tuple tuple) {
    ...
    if (srcStream .equals("s2")) {
      final String v = tuple.getStringByField("f2") + "&" +
                       tuple.getStringByField("f3") + "!!!";
      collector.emit(new Values(v));
      collector.ack(tuple);  ③
    } else {
      collector.emit(tuple,
                     new Values(tuple.getString(0) + "!!!"));
      collector.ack(tuple);  ④
    }
  }
}
```

① If any tuple is not acked within 10 s, it will be automatically failed.
② Enable at-least-once delivery semantics.
③④ We need to explicitly ack a tuple once it is successfully processed. A tuple can be failed by calling `collector.fail(tuple)`. If we do not explicitly invoke `ack()` or `fail()` on a tuple within the configured 10 s, which is set in the topology configuration, the spout will automatically fail the tuple.

The following code shows how to enable at-least-once during spout implementation:

```
import java.util.concurrent.ThreadLocalRandom;

public class TestWordSpout extends BaseRichSpout {
  public void nextTuple() {
    ...
    final long msgId1 = ThreadLocalRandom.current().nextLong();
    final long msgId2 = ThreadLocalRandom.current().nextLong();
    collector.emit("s1", v1, msgId1);  ①
    collector.emit("s2", v2, msgId2);  ②
    Logger.trace("emit {} {} {}", "s1", v1, msgId1);
    Logger.trace("emit {} {} {}", "s2", v2, msgId2);
    ...
  }
```

```
public void ack(Object msgId) { ③
  Logger.trace("ack: {}", msgId.toString());
}
public void fail(Object msgId) { ④
  Logger.trace("fail: {}", msgId.toString());
}
}
```

①②   We emit tuples with a message ID (msgId1 and msgId2 in this experiment), which is an Object. Each message needs to be annotated with
a unique ID, which enables the spout to track the messages. This ID is
supplied in ack() and fail() so that the user logic code knows which
tuple is acked or failed.

③   User logic code when a tuple is acked.

④   User logic code when a tuple is timeout or failed.

Follow the instructions in Sect. 5.2.2 to run the topology and check the log to
verify that the tuples are acked:

```
$ cd ~/heron-topology-java-at-least-once/ && mvn clean package

$ heron submit local ./target/heron-topology-java\
> -at-least-once-1.0-SNAPSHOT-jar-with-dependencies.jar \
> heronbook.ExclamationTopology \
> my-java-topology-at-least-once && \
> sleep 60 && tail -f /tmp/log.txt

$ heron kill local my-java-topology-at-least-once
```

6.1.2 Effectively-Once

When effectively-once was added to Heron, the initial requirements included:

Compatible with existing semantics
> The existing jobs with at-most-once and at-least-once should not be impacted in
> terms of both functionality and performance when we add effectively-once.

Negligible latency
> Effectively-once has to save states to some persistent storage, which could be
> time-consuming and impact latency. However, Heron is for real-time compu-
> tation, and we have to keep the latency overhead negligible compared to a no
> effectively-once setting.

Availability in all APIs
> To expose stateful processing and effectively-once features to topology writers,
> they should be provided through a low-level API. Higher-level APIs can leverage
> low-level APIs to implement advanced features, such as stateful window.

SPI for state storage
> Like other components in Heron, state stores should be implemented behind an
> abstraction SPI to align Heron extensibility.

6.1.2.1 Requirements for Effectively-Once

The topology must satisfy two prerequisites to be possibly effectively-once [1].

- It must be a stateful, idempotent topology.
- The input stream into the topology must be strongly consistent. In order to provide effectively-once semantics, a topology must be able to "rewind" state in case of failure. The state that it "rewinds" needs to be reliable—preferably durably stored.

  If the input to the topology is, for example, a messaging system that cannot ensure stream consistency, then effectively-once semantics cannot be applied, as the state "rewind" may return differing results. To put it somewhat differently, Heron can only provide delivery semantics as stringent as its data input sources can provide.

6.1.2.2 Exactly-Once Versus Effectively-Once

The Heron documentation reveals the story behind the name "effectively-once" and explains its difference from "exactly-once" [1]:

> There has been a lot of discussions recently surrounding so-called "exactly-once" processing semantics. We will avoid this term in the Heron documentation because we feel that it is misleading. "Exactly-once" semantics would mean that no processing step is ever performed more than once—and thus that no processing step is ever retried.
>
> It is important to always keep in mind that no system can provide exactly-once semantics in the face of failures (as this article argues). But that is okay because they do not need to; the truly important thing is that a stream processing system be able to recover from failures by "rewinding" state to a previous, pre-failure point and reattempt to apply processing logic. We use the term effectively-once, following Viktor Klang,[1] for this style of semantics.
>
> Heron can provide effectively-once guarantees if a topology meets the conditions outlined above, but it cannot provide "exactly-once" semantics.

6.1.2.3 Stateful Topologies

A topology can be identified as either stateful or nonstateful [1]:

- In stateful topologies, each component must implement an interface that requires it to store its state every time it processes a tuple (both spouts and bolts must do so).
- In nonstateful topologies, there is no requirement that any processing components store a state snapshot. Nonstateful topologies can provide at-most-once or at-least-once semantics, but never effectively-once semantics.

---

[1] https://twitter.com/viktorklang/status/789036133434978304.

For stateful topologies, there are two further types [1]:

- Idempotent stateful topologies are stateful topologies where applying the processing graph to an input more than once will continue to return the same result. A basic example is multiplying a number by 0. The first time you do so, the number will change (always to 0), but if you apply that transformation again and again, it will not change.

  For topologies to provide effectively-once semantics, they need to transform tuple inputs idempotently as well. If they do not, and applying the topology's processing graph multiple times yields different results, then effectively-once semantics cannot be achieved. If you would like to create idempotent stateful topologies, make sure to write tests to ensure that idempotency requirements are met.
- Nonidempotent stateful topologies are stateful topologies that do not apply processing logic along with the model of "multiply by zero" and thus cannot provide effectively-once semantics.

If you want to set the effectively-once semantic to a topology, it has to be both stateful and idempotent.

6.1.2.4 Implement Effectively-Once

To support effectively-once, a concept called `State` is added to each spout/bolt. `State` is essentially defined as a key-value map recording the computation results accumulated over time. When a spout/bolt implements the interface `IStatefulComponent`, its stateful processing feature is turned on. `IStatefulComponent` has two methods:

- `initState()`: When a spout/bolt runs for recovery, this method is invoked before spout `open()` and bolt `prepare()`. A single argument `state` is passed to the method, which contains the previously saved state from a checkpoint. In this method, the spout/bolt state should be restored according to the provided `state` argument. The spout/bolt should hold the `state` variable and save its state to this variable during the checkpointing process. When the topology is launched for the first time, an empty map is passed into the method through the `state` variable.
- `preSave()`: This method is invoked just before Heron saves the state during a checkpointing process, giving the user logic code a chance to perform some actions.

Take the Java topology in Sect. 5.2.1 as an example:

```
$ cp -r ~/heron-topology-java \
> ~/heron-topology-java-effectively-once
$ sed -i 's/-topology-java/-topology-java-effectively-once/g' \
> ~/heron-topology-java-effectively-once/pom.xml
```

We enable effectively-once in the topology configuration:

```java
public static void main(String[] args) throws Exception {
  ...
  Config conf = new Config();
  conf.setTopologyStatefulCheckpointIntervalSecs(20);  ①
  conf.setTopologyReliabilityMode(
    Config.TopologyReliabilityMode.EFFECTIVELY_ONCE);  ②
  conf.setDebug(true);
  ...
}
```

①   Set the checkpoint interval to 20 s.
②   Set the reliability mode to effectively-once.

In this example, we enable the state of the spout by inserting and updating the following code:

```java
import org.apache.heron.api.state.State;
import org.apache.heron.api.topology.IStatefulComponent;

public class TestWordSpout extends BaseRichSpout
    implements IStatefulComponent<String, Integer> {  ①
  private Map<String, Integer> myMap;  ②
  private State<String, Integer> myState;  ③

  ...

  @Override
  public void open(Map conf, TopologyContext context,
                   SpoutOutputCollector collector) {
    ...
    myMap = new HashMap<String, Integer>();
    for (String key: myState.keySet()) {
      myMap.put(key, myState.get(key));  ④
    }
  }

  public void nextTuple() {
    Utils.sleep(1000);
    final String[] words = new String[] {
      "nathan", "mike", "jackson", "golda", "bertels"};
    int offset = myMap.getOrDefault("current", 0);
    final String word = words[offset];
    myMap.put("current", ++offset % words.length);  ⑤
    collector.emit(new Values(word));
  }

  @Override
  public void initState(State<String, Integer> state) {  ⑥
    this.myState = state;  ⑦
  }
```

```
@Override
public void preSave(String checkpointId) {  (8)
  for (String key : myMap.keySet()) {
    myState.put(key, myMap.get(key));  (9)
  }
 }
}
```

(1)   We let spout implement the IStatefulComponent interface. We changed
       the random emit word to sequence emit, and the state is the index on the
       word array indicating the current emit position.

(2)   A member variable to hold the stream index position. It is the state of the
       spout, and we separate it from myState fed in initState() to emulate
       a spout with states in multiple members so that its state needs unpack in
       initState() and pack in preSave().

(3)   Hold the state fed in initState().

(4)   Unpack state to the spout member. initState() is called before open();
       thus, the member, myState, is initialized and ready to populate the other
       members in open().

(5)   Update the spout state.

(6)   initState() passes the state variable to the spout, which should be held to
       store its state.

(7)   Hold the state variable for use in open() and preSave().

(8)   preSave() is called just before the checkpoint operation. The spout has the
       chance to populate or update the state that will be checkpointed.

(9)   Populate the spout state to myState for checkpointing.

Run the topology, kill one Heron Instance process, and then watch how the state
is restored when the process comes back:

```
$ cd ~/heron-topology-java-effectively-once/ && mvn clean package

$ heron submit local target/heron-topology-java-\
> effectively-once-1.0-SNAPSHOT-jar-with-dependencies.jar \
> heronbook.ExclamationTopology \
> my-java-topology-effectively-once && \
> sleep 60 && tail -f /tmp/log.txt  (1)

$ kill `ps -ef | grep HeronInstance | grep java | \
> awk '{print $2}' | head -1`  (2)

$ heron kill local my-java-topology-effectively-once
```

(1)   Watch the log after the topology is running.

(2)   Kill one Heron Instance process in a separate SSH session (there should
       be six Heron Instance processes in total) and watch the log showing that it
       restores state.

6.2   Windowing

Heron has support for grouping tuples within a window and processing them in one batch. Heron windowing is implemented in the Heron API by introducing a set of bolt interfaces and classes, among which two bolt interfaces and their corresponding implementation classes are essential:

- `IWindowedBolt` and `BaseWindowedBolt`: Class `BaseWindowedBolt` implements interface `IWindowedBolt`. `IWindowedBolt` is similar to the other bolt interfaces except that its `execute()` accepts `TupleWindow` instead of `Tuple`. `TupleWindow` represents a list of tuples in a window.
- `IStatefulWindowedBolt` and `BaseStatefulWindowedBolt`: These integrate windowing to stateful bolt.

6.2.1   Windowing Concepts

Windows can be specified with two parameters:

Window length   the window length or duration.
Sliding interval   the interval at which the windowing slides.

Both window length and sliding interval can be measured by either a tuple count or a time duration.

The time when the bolt processes the tuple, called processing timestamp, is used by window calculations by default. Another often-used timestamp called event timestamp is the source-generated timestamp, which Heron also supports through `withTimestampField(fieldName)`. `fieldName` indicates the timestamp field in the tuple, which will be considered for windowing calculations. An exception will be thrown if the specified field is not found. More generally, a `TimestampExtractor` object can be adopted to extract a timestamp from a tuple by `withTimestampExtractor(timestampExtractor)`.

When using event timestamp, we can specify a time lag parameter to indicate the maximum out of order time limit for tuples by `withLag(duration)`. This behavior can be changed by specifying a stream ID in `withLateTupleStream(streamId)`. In this case, the specified stream of `streamId` will hold late tuples accessible through the `WindowedBoltExecutor.LATE_TUPLE_FIELD` field.

The concept "watermark" was introduced to differentiate window data and late data. It is essentially a timestamp, and Heron calculates watermarks based on the event timestamp of input streams. A watermark is calculated by

$$min_j(max_i(timestamp_{i,j})) - lag \tag{6.1}$$

where $i \in tuples$ and $j \in streams$, which is the minimum min_j of the latest tuple timestamps $max_i(timestamp_{i,j})$ across all input streams minus the lag. Apache

Beam[2] has a similar watermark concept. The watermark timestamps are emitted periodically (the default is every second), and it is the clock tick for the window calculation for event timestamp. When a watermark arrives, all windows till that watermark are evaluated.

6.2.2 Windowing Example

To enable windowing, you can simply inherit IWindowedBolt. Take the topology in Sect. 5.2.1 as an example:

```
$ cp -r ~/heron-topology-java ~/heron-topology-java-window
$ sed -i 's/heron-topology-java/heron-topology-java-window/g' \
> ~/heron-topology-java-window/pom.xml
```

The following code enables windowing in the file *ExclamationTopology.java*:

```
import org.apache.heron.api.bolt.BaseWindowedBolt;
import org.apache.heron.api.windowing.TupleWindow;
import java.util.stream.Collectors;

public static class ExclamationBolt
    extends BaseWindowedBolt { ①
  public void execute(TupleWindow inputWindow) { ②
    String output = inputWindow.get().stream() ③
      .map(tuple -> {
        String srcStream = tuple.getSourceStreamId();
        if ("s2".equals(srcStream)) {
          return tuple.getStringByField("f2") + "&" +
                 tuple.getStringByField("f3");
        } else {
          return tuple.getString(0);
        }})
      .distinct().sorted().collect(Collectors.joining(","));
    Logger.trace("emit {}", output);
    collector.emit(new Values(output));
  }
}


public static void main(String[] args) throws Exception {
  ...
  builder.setBolt("exclaim1", new ExclamationBolt()
    .withTumblingWindow(BaseWindowedBolt.Count.of(10)), 2) ④
        .shuffleGrouping("word", "s1")
        .shuffleGrouping("word", "s2");
  builder.setBolt("exclaim2", new ExclamationBolt()
    .withTumblingWindow(BaseWindowedBolt.Count.of(10)), 2)
        .shuffleGrouping("exclaim1");
  ...
}
```

---

[2]https://beam.apache.org/.

①   The bolt class extends `BaseWindowedBolt`.
②   The argument of `execute()` is `TupleWindow`.
③   `TupleWindow.get()` returns a `List` of `Tuple` within the configured window.
④   Set the window configuration on the bolt.

To enable lambda expressions, we need Java 8 or higher and thus set the source code to Java 8 in *pom.xml*:

```
<properties>
  <maven.compiler.source>1.8</maven.compiler.source>
  <maven.compiler.target>1.8</maven.compiler.target>
</properties>
```

Compile, run the topology, and watch the logs:

```
$ cd ~/heron-topology-java-window/ && mvn clean package

$ heron submit local target/heron-topology-java\
> -window-1.0-SNAPSHOT-jar-with-dependencies.jar \
> heronbook.ExclamationTopology my-java-topology-window && \
> sleep 60 && tail -f /tmp/log.txt

$ heron kill local my-java-topology-window
```

6.3 Summary

In this chapter, we studied Heron delivery semantics, including at-most-once, at-least-once, and effectively-once, and tried to enable the topology to at-least-once by acking and effectively-once by stateful processing.

Up until now, we have become familiar with the low-level Topology API. In the next chapter, we will see the Streamlet API inspired by functional programming. The Streamlet API changes the directed acyclic topology model to a stream processing graph.

Reference

1. Heron delivery semantics. https://heron.incubator.apache.org/docs/heron-delivery-semantics. Visited on 2020-07

Chapter 7
Heron Streamlet API

When Heron was born, it came with a Topology API, which was quite similar to the Storm API due to the goal of being compatible with the Storm API. Like the Storm API, the Heron Topology API requires topology writers to describe the topology structure explicitly to:

- Define the DAG vertices, or spout/bolt, including setup, loop of tuple operations, and teardown behaviors
- Define the DAG edges, or streams, including the pair of vertices and how the stream is partitioned

The Heron Topology API has been proven to be working well for all kinds of topologies in production. However, a few of its drawbacks are as follows [2]:

Verbosity

In the original Topology API for both Java and Python, creating spouts and bolts required substantial boilerplate. This forced developers to both provide implementations for spout and bolt classes and specify the connections between those spouts and bolts.

Difficult debugging

When spouts, bolts, and the connections between them need to be created "by hand," it can be challenging to trace the origin of problems in the topology's processing chain.

Tuple-based data model

In the older Topology API, spouts and bolts passed tuples and nothing but tuples within topologies. Although tuple is a powerful and flexible data type, the Topology API forced all spouts and bolts to implement their own serialization/deserialization logic.

H. Wu, M. Fu, *Heron Streaming*, https://doi.org/10.1007/978-3-030-60094-5_7

To overcome these complaints as well as to align with the functional style programming trend, Heron provides a Streamlet API, with which you do not have to implement spout/bolt or connecting vertices. More specifically, the Heron Streamlet API offers [2]:

Boilerplate-free code
   Instead of needing to implement spout and bolt classes over and over again, the Heron Streamlet API enables you to create stream processing logic out of functions, such as map, flatMap, join, and filter functions.
Easy debugging
   With the Heron Streamlet API, you do not have to worry about spouts and bolts, which means that you can more easily surface problems with your processing logic.
Completely flexible, type-safe data model
   Instead of requiring that all processing components pass tuples to one another (which implicitly requires serialization to and deserialization from your application-specific types), the Heron Streamlet API enables you to write your processing logic in accordance with whatever types you would like—including tuples, if you wish. In the Streamlet API for Java, all streamlets are typed (e.g., Streamlet<MyApplicationType>), which means that type errors can be caught at compile time rather than at runtime.

Table 7.1 [1] lists some crucial differences between the two APIs. While the Streamlet API provides a new data model for programming, it shares a few common things with the Topology API:

- Both the Streamlet API and the Topology API are used to describe the topology logical plans. The Heron Streamlet API internally still creates topologies but is hidden from topology writers. Usually, Sources in the Streamlet API are converted to Spout, and the other operations and Sinks are converted to Bolt. Although the difference is obvious when composing topologies, once the logical plan is determined, the packing and physical plans are handled in the same way by Heron automatically. To schedulers/containers, ZooKeeper, and shared package storages, the Streamlet API and the Topology API are no different. Eventually, you can regard the Streamlet API as a high-level wrapper on the Topology API with a new data model called "processing graph."
- The APIs are for topology writers, and they are transparent for topology operators who use the same Heron CLI to manage both. In other words, both APIs are equivalent to SREs.

The Streamlet API is encouraged for most application scenarios, which frees efforts on DAG details and lets you focus on business logic. It is in functional style that fits more smoothly with stream concepts. On the contrary, the Topology API can be adopted when you are going to control every corner of the DAG. The fine-grained control makes the job tuning more effective, especially for critical jobs. Overall, you have the choice to pick either API. The jobs of different APIs can coexist and perform well in the same environment.

Table 7.1 The Heron Streamlet API versus the Topology API

Domain	Original Topology API	Heron Streamlet API
Programming style	Procedural, processing component based	Streamlet
Abstraction level	Low level. Developers must think in terms of "physical" spout and bolt implementation logic	High level. Developers can write processing logic in an idiomatic fashion in the language of their choice, without needing to write and connect spouts and bolts
Processing model	Spout and bolt logic must be created explicitly, and it is the developer's responsibility to connect the spouts and bolts	Spouts and bolts are created automatically based on the processing graph that you build

Fig. 7.1 Heron Streamlet general model

7.1 Streamlet API Concepts

Unlike the Topology API of spouts and bolts, the Streamlet API introduced the notion of a "processing graph," which consists of the following components [2]:

Sources   supply the processing graph with data from random generators, databases, web service APIs, file systems, pub-sub messaging systems, or anything that implements the source interface.

Operators   supply the graph's processing logic, operating on data passed into the graph by sources.

Sinks   are the terminal endpoints of the processing graph, determining what the graph does with the processed data. Sinks can involve storing data in a database, logging results to stdout, publishing messages to a topic in a pub-sub messaging system, and much more.

Connecting the three components, a general stream looks like Fig. 7.1.

7.1.1   Streamlets

The core concept of the Heron Streamlet API is "streamlet." A processing graph is made of streamlets. A streamlet is an unbounded sequence of tuples flowing from the source to the sink. A streamlet can split or diverge into multiple streamlets by a Clone operation, and multiple streamlets can merge into a single streamlet by a union or join operation. Operations can be applied to the streamlet in a sequence.

Like the Topology API, the source and sink of a streamlet could be messaging systems like Apache Pulsar[1] and Google Cloud Pub-Sub, file readers/writers, random generators (for source), and storages like Apache Druid and Google BigQuery (for sink). Unlike the Topology API, the operators on streamlets are more like functional programming, including map/flatMap, reduce, etc. Besides the tuples and operators, a streamlet has the following properties associated with it [3]:

name
> The user-assigned or system-generated name to refer to the streamlet.

nPartitions
> The number of partitions that the streamlet is composed of. Thus the ordering of the tuples in a streamlet is for tuples within a partition. This allows the system to distribute each partition to different nodes across the cluster.

When you apply an operator on the streamlet, you apply the operator on each tuple of that streamlet. You can think of a streamlet as a vehicle fleet on the highway, and the operators are tolls. The streamlet partition is like the lane, but a tuple can change lanes at tolls. Each toll booth is for a single lane; thus, the operator parallelism is the same as the streamlet partition count. If you want to adjust the operator parallelism, you have to change the stream partition count by repartitioning the streamlet.

7.1.2   Operations

The Heron Streamlet API supports a variety of operators to transform streamlets. These operators are inspired by functional programming, and you will find them familiar just as though you were doing functional programming. These operators are summarized in Table 7.2 [2].

---

[1] https://pulsar.apache.org/.

Table 7.2 Streamlet operations

Operation	Description	Example
map	Create a new streamlet by applying the supplied mapping function to each element in the original streamlet	Add 1 to each element in a streamlet of integers
flatMap	Like a map operation but with the important difference that each element of the streamlet is flattened	Flatten a sentence into individual words
filter	Create a new streamlet containing only the elements that satisfy the supplied filtering function	Remove all inappropriate words from a streamlet of strings
union	Unify two streamlets into one, without modifying the elements of the two streamlets	Unite two different Streamlet<String>s into a single streamlet
clone	Create any number of identical copies of a streamlet	Create three separate streamlets from the same source
transform	Transform a streamlet using whichever logic you would like (useful for transformations that do not neatly map onto the available operations)	
join	Create a new streamlet by combining two separate key-value streamlets into one based on each element's key	Combine key-value pairs listing current scores (e.g., ("h4x0r", 127)) for each user into a single per-user stream
reduceByKey-AndWindow	Produce a streamlet out of two separate key-value streamlets on a key, within a time window, and by a reduce function that you apply to all the accumulated values	Count the number of times a value has been encountered within a specified time window
repartition	Create a new streamlet by applying a new parallelism level to the original streamlet	Increase the parallelism of a streamlet from 5 to 10
toSink	Sink operations terminate the processing graph by storing elements in a database, logging elements to stdout, etc.	Store processing graph results in an AWS Redshift table
log	Log the final results of a processing graph to stdout; this must be the last step in the graph	
consume	Consume operations are like sink operations except they do not require implementing a full sink interface (consume operations are thus suited for simple operations like logging)	Log processing graph results using a custom formatting function

The following sections will demonstrate the streamlet operations using examples.

7.2 Write a Processing Graph with the Java Streamlet API

Since the Streamlet API eventually transforms the streamlets and processing graph
to a topology, let us revisit the example topology in Sect. 5.1.1 and see its equivalent
processing graph code in `ExclamationTopology.java:main()`:

```
$ cp -r ~/heron-topology-java ~/heron-streamlet-java && \
> cd ~/heron-streamlet-java
$ sed -i 's/heron-topology-java/heron-streamlet-java/g' pom.xml
$ rm src/main/java/heronbook/TestWordSpout.java
```

Remove `class ExclamationBolt` in *ExclamationTopology.java*. Update `main()`
to:

```
public static void main(String[] args) throws Exception {
  String[] words = new String[] {
    "nathan", "mike", "jackson", "golda", "bertels"};

  Builder builder = Builder.newBuilder();  ①

  Streamlet<String> s1 = builder.newSource(() -> {
    Utils.sleep(500);
    return words[ThreadLocalRandom
                 .current().nextInt(words.length)];
  }).setName("s1");  ②
  Streamlet<String[]> s2 = builder.newSource(() -> {
    Utils.sleep(500);
    String word2 =
      words[ThreadLocalRandom.current().nextInt(words.length)];
    String word3 =
      words[ThreadLocalRandom.current().nextInt(words.length)];
    return new String[] {word2, word3};
  }).setName("s2");  ③

  Streamlet<String> s3 = s1.map(x -> x+" !!!");  ④
  Streamlet<String> s4 = s2.map(x -> x[0]+" & "+x[1]+" !!!");  ⑤
  s3.union(s4).map(x -> x+" !!!").log();  ⑥

  Config config = Config.newBuilder().setNumContainers(2)
                        .build();  ⑦

  new Runner().run("my-java-streamlet", config, builder);  ⑧
}
```

①   Initialize the builder to create the processing graph. Each builder object
handles one single graph, which is required by the submission function.

(2)   The s1 source streamlet emits one word per tuple. Each source generates
          a single stream, which is different from the spout that can emit multiple
          streams.

(3)   The s2 source streamlet emits two words per tuple. The streamlet source
          does not declare fields; thus, all fields should be encapsulated in some data
          structure.

(4)   A map operator emulates the bolt by adding "!!!" to the word.

(5)   A map operator emulates the bolt by joining two words and appending "!!!"
          at the end.

(6)   Union two streams and add another "!!!", then log the content, emulating
          the terminal bolt operations.

(7)   Construct the configuration for the whole processing graph. In this example,
          set the container count to 2.

(8)   Submit the processing graph, where `my-java-streamlet` is the processing
          graph name.

The above code showed the three general steps to write a processing graph: building
the graph with `Builder`, building the `Config` object, and submitting the graph
with `Runner`. This example code demonstrates only a small part of the powerful
Streamlet API, while the following subsections will describe the Streamlet API in
detail with example code by categories.

  Follow the similar steps in Sect. 6.2.2 to update the *pom.xml* file, and then update
`imports` in the Java file. Compile and run this processing graph as below. `LogSink`
will print the tuples in its log file.

```
$ cd ~/heron-streamlet-java/ && mvn clean package

$ heron submit local target/\
> heron-streamlet-java-1.0-SNAPSHOT-jar-with-dependencies.jar \
> heronbook.ExclamationTopology my-java-streamlet

$ heron kill local my-java-streamlet
```

7.2.1   Sources

In this subsection, we will see how to define two streamlets, *s1* and *s2*, by two
generation approaches; they will be used in the rest of the Java example code to
demonstrate the operator usage. There are two sources in the Streamlet API, the
simpler one being derived from `SerializableSupplier`, an extension of the Java
`Serializable` and `Supplier` interfaces. `Supplier` generates a tuple per invocation
and can be written as a Java lambda expression, as shown in the following example,
generating a random character in the range from "A" to "D":

```
Streamlet<Character> s1 = builder
  .newSource(() ->
    (char)ThreadLocalRandom.current().nextInt('A', 'D'))
  .setName("SerializableSupplier-ABCD");
```

The other source type is `Source`. The `get()` method invocation returns a
new element that forms the tuples of the streamlet. `setup()` and `cleanup()` are
where the generator can finish any one-time setup work, like establishing/closing
connections to sources, etc. The following code shows how to generate a random
character stream whose element is in the range from "C" to "F":

```
Streamlet<Character> s2 = builder
  .newSource(new Source<Character>() {
    @Override
    public void setup(Context context) {}

    @Override
    public Collection<Character> get() {
      return ThreadLocalRandom.current().ints(7, 'C', 'F')
        .mapToObj(x -> (char)x).collect(Collectors.toList());
    }

    @Override
    public void cleanup() {}
  })
  .setName("Source-CDEF");
```

The `Source` type is more flexible than `SerializableSupplier` in two aspects.
First, it generates a batch of tuples per invocation. Second, it allows preparation and
clean tasks once before and after the tuples generation. `Source` can do the same
tasks `SerializableSupplier` does. For example, to generate the same streamlet
with the above `s1`:

```
Streamlet<Character> s1_ = builder
  .newSource(new Source<Character>() {
    @Override
    public void setup(Context context) {}

    @Override
    public Collection<Character> get() {
      return Arrays.asList(
        (char)ThreadLocalRandom.current().nextInt('A', 'D'));
    }

    @Override
    public void cleanup() {}
  })
  .setName("Source-ABCD");
```

7.2.2   *Sinks*

The sink is the end of a stream; in other words, it does not return a streamlet as sources and operators do. It has one input and zero output, and no more operators can be chained after a sink. The streamlet flows into it and does not come back from a processing graph perspective. However, the data may be sent to external systems like being published on a messaging system or written on a database.

There are three sinks in the Streamlet API. The most simple sink is `log()`, which writes the tuples in the streamlet into log files using the `String.valueOf()` function. The other two sinks are similar to the two sources. `consume()` accepts a `SerializableConsumer` parameter as well as a Java lambda expression, while `toSink()` accepts a `Sink` parameter that allows setup and cleanup operations. The following code shows three sinks to implement the same task—log every tuple in the streamlet:

```
s1.log();
s1.consume(tuple -> System.out.println(String.valueOf(tuple)));
s1.toSink(new Sink<Character>() {
  @Override
  public void setup(Context context) {}

  @Override
  public void put(Character tuple) {
    System.out.println(String.valueOf(tuple));
  }

  @Override
  public void cleanup() {}
});
```

7.2.3   *Transform: Filter, Map, FlatMap*

The `filter()`, `map()`, and `flatMap()` operators are categorized together because they map tuple to tuple. `filter()` inspects the tuple, keeps the useful tuple, and removes the rest. The processing graph writer provides the inspection function through a `SerializablePredicate` parameter or a lambda expression. `map()` maps a single tuple to another single tuple. The input and output tuples may be of different types. `flatMap()` applies a function, which maps one tuple to multiple tuples, to each element of the streamlet.

Besides these three operators, a general operator `transform()` is available to fulfill operations that do not neatly fit into the previous three categories. It is also useful for operations involving any states in stateful topologies. `transform()` requires you to implement three different methods [1]:

Table 7.3 Comparison of `filter()`, `map()`, `flatMap()`, and `transform()`

	`filter()`	`map()`	`flatMap()`	`transform()`
tuple count	less	same	more	any
tuple type	same	may change	may change	may change

- A `setup()` method that enables you to pass a `Context` object to the operation and to specify what happens before the `transform()` step. The `Context` object available to a transform operation provides access to the current state of the processing graph, the processing graph's configuration, the name of the stream, the stream partition, and the current task ID.
- A `transform()` operation that performs the desired transformation.
- A `cleanup()` method that allows you to specify what happens after the `transform()` step.

The four operator features are summarized in Table 7.3.

The following code shows how to keep "A" and "C" in the stream using `filter()` and equivalent `transform()`:

```
Streamlet<Character> f1 =
  s1.filter(x -> x.equals('A') || x.equals('C'));
Streamlet<Character> t1 = s1.transform(
    new SerializableTransformer<Character, Character>() {
  @Override
  public void setup(Context context) {}

  @Override
  public void transform(Character x,
                        Consumer<Character> consumer) {
    if (x.equals('A') || x.equals('C')) {
      consumer.accept(x);
    }
  }

  @Override
  public void cleanup() {}
});
```

The following code shows how to map "A" to "true" and "C" to "false" in the stream using `map()` and equivalent `transform()`:

```
Streamlet<Boolean> m1 =
  f1.map(x -> x.equals('A') ? true : false);
Streamlet<Boolean> t2 = t1.transform(
    new SerializableTransformer<Character, Boolean>() {
  @Override
  public void setup(Context context) {}

  @Override
  public void transform(Character x,
                        Consumer<Boolean> consumer) {
```

```
    if (x.equals('A')) {
      consumer.accept(true);
    } else {
      consumer.accept(false);
    }
  }

  @Override
  public void cleanup() {}
});
```

The following code shows how to flatten "true" to "A," "B" and "false" to "C,"
"D" in the stream using flatMap() and equivalent transform(). The order of the
flattened tuples is kept in the final stream; thus "B" always follows "A," and "D"
always follows "C":

```
Streamlet<Character> fm1 = m1.flatMap(
    x -> x ? Arrays.asList('A', 'B') : Arrays.asList('C', 'D'));
Streamlet<Character> t3 = t2.transform(
    new SerializableTransformer<Boolean, Character>() {
  @Override
  public void setup(Context context) {}

  @Override
  public void transform(Boolean x,
                        Consumer<Character> consumer) {
    if (x) {
      consumer.accept('A');
      consumer.accept('B');
    } else {
      consumer.accept('C');
      consumer.accept('D');
    }
  }

  @Override
  public void cleanup() {}
});
```

The streamlet operator functions can be chained since they return a new
Streamlet object. The chain style of the above example looks like this:

```
s1.filter(x -> x.equals('A') || x.equals('C'))
  .map(x -> x.equals('A') ? true : false)
  .flatMap(x -> x ? Arrays.asList('A', 'B')
                  : Arrays.asList('C', 'D'));
```

7.2.4   Partitioning

For the Topology API, you can set the parallelism of spouts and bolts, but there are
no configurations for partitions of streams. For the Streamlet API, you do not set

the parallelisms on operators; instead, you set the partition count on the streams. Keep in mind that the primary construct in processing graphs is the streamlet, and the operator parallelism follows the streamlet partition count.

The default streamlet partition count is 1. When the streaming throughput is high, you can increase the partition count, called fan-out; on the other side, you can reduce the partition count, called fan-in. The following example code sets the spout parallelism to 1:

```
s1.setNumPartitions(1);
```

Once you set the streamlet partition, the following streamlets keep the same partition, and operators get the same parallelism. You have to explicitly set a new partition count when you would like to update it. To change the streamlet partition count, add the operator `repartition(nPartitions)` to the streamlet. It is internally a `filter` operator followed by a `setNumPartitions()` invocation—`filter(filterFn).setNumPartitions(nPartitions)`, where `filterFn` is the identity function.

Besides simply changing the partition count, you can also take a fine-grained control to determine which particular partition a tuple should go to through `repartition(nPartitions, partitionFn)`. `nPartitions` indicates how many partitions you want, and `partitionFn` is a function that users should implement to indicate how a tuple is routed to partitions. More specifically, `partitionFn` accepts two arguments—the tuple and `nPartitions`—and returns a list of partition indexes each in the range $[0, nPartitions)]$. Its return list could be empty, which means "do not pass this tuple to downstream," or include multiple numbers, which means "emit this tuple to multiple partitions." This example shows this advanced control on the repartitioning:

```
s1.repartition(3, (c, i) -> {
  return Arrays.asList(c%2);
});
```

The above `repartition()` example sets the bolt parallelism to 3. However, it routes the tuples to only the first and second bolt instances while letting the third bolt instance idle, which is for demonstration purpose—`repartition()` can adjust grouping.

7.2.5 Clone and Union

The clone operator creates copies of a streamlet. Each copy streamlet is exactly the same as the original streamlet. The copy is a deep copy; in other words, the copies do not impact each other and do not impact the original streamlet. From a topology perspective, the bolt of the original streamlet emits to n bolts that represent the n copied streamlets; in other words, $(1 + n)$ bolts together implement a "1 to n" clone operation.

The union operator simply puts the tuples in two streamlets into one streamlet and keeps their order. The tuple types of the two streamlets into the union operator should be the same, and it is the tuple type of the final combined streamlet. From a topology perspective, two components emit tuples to a single component; in other words, (2 + 1) bolts together implement a "2 to 1" union operation.

The following example code shows how to clone and union streamlets:

```
List<Streamlet<Character>> cl = s1.clone(2);
Streamlet<Character> ul = s1.union(s2);
```

7.2.6 Reduce by Key and Window

A reduce operator is specified by the following arguments [1]:

- A time window across which the operation will take place.
- A key extractor that determines what counts as the key for the streamlet.
- A value extractor that determines which final value is chosen for each element of the streamlet.
- A reduce/aggregation function that produces a single value for each key in the streamlet. There are two reduce function types: reduce two values to one and reduce each value against an initial value.

The "Reduce by key and window" operator accepts one streamlet and produces a new streamlet. The operator first classifies tuples according to two dimensions—window and key—then applies the reduce function on the tuples in each set. The window dimension in the classification triggers the output for an unbounded sequence of tuples. For bounded streams, you can include all the tuples in a single window to emulate the batch processing. The key dimension in the classification is introduced because we often need to group data before applying aggregation in the real world. If you do not need this further classification, use the same key for the key extractor function. Each tuple in the output streamlet is a key-value pair, and the key is a two-element list <key, window> encapsulated in a `KeyedWindow` object. In simple words, each tuple in the output streamlet is: <<key, window> → value>. For each `KeyedWindow` object corresponding to an individual tuple set from the classification, there is a single reduced value and thus a single tuple.

Here are examples to count the character frequency within windows:

```
KVStreamlet<KeyedWindow<Character>, Integer> r1 = s1
  .reduceByKeyAndWindow(x -> x, x -> 1,   ①
    WindowConfig.TumblingCountWindow(5),   ②
    (x, y) -> x + y);   ③
KVStreamlet<KeyedWindow<Character>, Integer> r2 = s2
  .reduceByKeyAndWindow(x -> x,   ④
    WindowConfig.SlidingTimeWindow(
      Duration.ofSeconds(3), Duration.ofSeconds(1)),   ⑤
    0, (x, y) -> x + 1);   ⑥
```

①   Use the character itself as the key and set the frequency value to 1 for each
      character in the stream.
②   A count tumbling window of five tuples. The `Window` object has
      two members: count and time range. It can be of either a count
      window or a time window. `WindowConfig` defines the window type and
      its properties. Currently, Heron supports four predefined windows—
      `TumblingTimeWindow`, `SlidingTimeWindow`, `TumblingCountWindow`,
      and `SlidingCountWindow`. You can also define your window through
      `CustomWindow`.
③   Aggregate the values by adding two frequencies of the same key. Both x and
      y represent the frequency values.
④   Key extraction function. For this type of reduction, no value extraction
      function is necessary since the values are calculated in the reduction
      function.
⑤   A time sliding window of size 3 s triggered each second.
⑥   The initial identity value is 0. x represents the frequency value and y
      represents the tuple. The reduce function adds one against the frequency
      value for each character in the window of the same key.

7.2.7   Join

The join operator accepts two streamlets (a "left" and a "right" streamlet) and
outputs a single streamlet [2]:

- Based on a key extractor for each streamlet
- Over key-value elements accumulated during a specified time window
- Based on a join type (inner, outer left, outer right, or outer)
- Using a join function that specifies how values will be processed

All join operations are performed [2]:

- Over elements accumulated during a specified time window
- Per a key and value extracted from each streamlet element (you must provide
  extractor functions for both)
- By a join function that produces a "joined" value for each pair of streamlet
  elements

The following code shows how to do an outer left join on streams r1 and r2:

```
KVStreamlet<KeyedWindow<Character>, Integer> j2 = r1.join(
    r2, x -> x.getKey().getKey(), y -> y.getKey().getKey(), ①
    WindowConfig.TumblingCountWindow(11), JoinType.OUTER_LEFT, ②
    (x, y) -> (x==null ? 0 : x.getValue()) +
              (y==null ? 0 : y.getValue())); ③
```

①   Use the key inside KeyedWindow as the stream key for both streams.
②   Indicate the window and one of the four join types.
③   The join function adds two frequency values together. For the outer joins, the join function has to handle null since there may be unmatched keys.

Comparing `join` and `reduceByKeyAndWindow`, there are two differences. First, `join` needs a `JoinType` parameter, which determines how we merge keys from two streamlets into keys in a single streamlet. The output tuple count depends on the final key count or the `JoinType` argument. Second, `join` replaces the reduction/aggregation function to a join function, which first extracts values from two tuples from left and right streamlets, respectively, and reduces/aggregates them into a single value.

The Heron Streamlet API supports four types of joins, as listed in Table 7.4 [2]. Inner joins operate over the Cartesian product of the left stream and the right stream. Imagine this set of key-value pairs accumulated within a time window, shown in Table 7.5 [2]. An inner join operation would thus apply the join function to all key values with matching keys, producing this set of key values with the join function adding the values together, as shown in Table 7.6 [2].

An outer left join includes the results of an inner join plus all of the unmatched keys in the left stream. The resulting set of key values within the time window is shown in Table 7.7 [2]. An outer right join includes the results of an inner join plus all of the unmatched keys in the right stream, as shown in Table 7.8 [2]. Outer joins include all key values across both the left and right streams, regardless of whether or not any given element has a matched key in the other stream. If you want to ensure

Table 7.4 Join types

Type	Default	What the join operation yields
Inner	Yes	All key values with matched keys across the left and right streams
Outer left	No	All key values with matched keys across both streams, plus unmatched keys in the left stream
Outer right	No	All key values with matched keys across both streams, plus unmatched keys in the right stream
Outer	No	All key values across both the left and right streams, regardless of whether or not any given element has a matching key in the other stream

Table 7.5 Input streams

Key	Value	Key	Value
A	1	C	6
B	2	C	7
C	3	D	8
D	4	E	9
D	5	F	10

Table 7.6 Inner
join

Key	Value
C	3 + 6 = 9
C	3 + 7 = 10
D	4 + 8 = 12
D	5 + 8 = 13

Table 7.7 Outer
left join

Key	Value
C	3 + 6 = 9
C	3 + 7 = 10
D	4 + 8 = 12
D	5 + 8 = 13
A	1 + null = 1
B	2 + null = 2

Table 7.8 Outer
right join

Key	Value
C	3 + 6 = 9
C	3 + 7 = 10
D	4 + 8 = 12
D	5 + 8 = 13
E	null + 9 = 9
F	null + 10 = 10

that no element is left out of a resulting joined streamlet, use an outer join, as shown
in Table 7.9 [2].

7.2.8   Configuration

Besides the processing graph structure, configurations are also necessary to shape
the processing graph. The streamlet configuration processes graphwise and is
selected from the most often used items in the Topology API. A summary of the
available configurations is shown in Table 7.10.

Table 7.9 Outer join

Key	Value
C	3 + 6 = 9
C	3 + 7 = 10
D	4 + 8 = 12
D	5 + 8 = 13
A	1 + null = 1
B	2 + null = 2
E	null + 9 = 9
F	null + 10 = 10

Table 7.10 Streamlet configurations

Category	Configurable items
NumContainers	count
PerContainer resources	CPU, RAM
DeliverySemantics	ReliabilityMode default ATMOST_ONCE
Serializer	default KRYO
UserConfig	key-value pair

7.3 Write a Processing Graph with the Python Streamlet API

In this section, we will make an equivalent Python streamlet processing graph to the example topology in Sect. 5.1.1. We follow the same steps in Sect. 5.3 to establish a project directory. The code directory looks like:

```
$ cp -r ~/heron-topology-py/ ~/heron-streamlet-py/ && \
> cd ~/heron-streamlet-py
$ ./pants clean-all && rm -rf dist/
$ rm spout.py bolt.py
$ > topology.py

$ tree --charset=ascii ~/heron-streamlet-py/
/home/ubuntu/heron-streamlet-py/
|-- BUILD
|-- heronpy-0.0.0-py3-none-any.whl
|-- pants
|-- pants.toml
`-- topology.py
```

7.3.1   Source Generator

topology.py includes a parent class defining the candidate word set. The two spout streams are implemented by two `Generators` with `sleep()` to slow down emitting speed:

```
from random import choice
from time import sleep

from heronpy.streamlet.generator import Generator  ①

class SlowGenerator(Generator):  ②
  def setup(self, context):  ③
    self._words = (
      'nathan', 'mike', 'jackson', 'golda', 'bertels')

class SlowGenerator1(SlowGenerator):
  def get(self):
    sleep(1)  ④
    return choice(self._words)  ⑤

class SlowGenerator2(SlowGenerator):
  def get(self):
    sleep(1.5)
    return (choice(self._words), choice(self._words))  ⑥
```

①   The `Generator` source.
②   Inherit `Generator`.
③   Implement `setup()`.
④   Delay making source slow for easy observation.
⑤   Pick one word randomly.
⑥   Construct a tuple with two words.

7.3.2   Processing Graph Construction

topology.py includes the following processing graph construction:

```
from heronpy.streamlet.builder import Builder
from heronpy.streamlet.config import Config
from heronpy.streamlet.runner import Runner

if __name__ == '__main__':
  builder = Builder()  ①

  stream_1 = builder.new_source(SlowGenerator1())  ②
  stream_2 = builder.new_source(SlowGenerator2())  ③

  stream_3 = stream_1.map(lambda x: x + " !!!")  ④
  stream_4 = stream_2 \
```

```
    .map(lambda x: x[0] + " & " + x[1] + " !!!")  ⑤

stream_3.union(stream_4).map(lambda x: x + " !!!").log()  ⑥

Runner().run("my-python-streamlet", Config(), builder)  ⑦
```

① The builder to draw the processing graph.
② The new source of a one-word tuple.
③ The new source of a two-word tuple.
④ Append "!!!" to the one-word tuple.
⑤ Concatenate two words then append "!!!".
⑥ Union two streams, then append "!!!".
⑦ Use Runner to submit the processing graph with the default configuration.

The following commands show how to compile and submit the job:

```
$ cd ~/heron-streamlet-py && \
> ./pants clean-all && ./pants binary //:

$ heron submit local dist/exclamation-topology.pex -

$ heron kill local my-python-streamlet
```

7.4 Write a Processing Graph with the Scala Streamlet API

The Scala API is for streamlets, particularly. There is no Scala API for low-level
APIs. In this section, we first establish a Scala development environment, then
construct the project source structure, and then compose the Scala code to build
a streamlet processing graph.

7.4.1 Install sbt

To manage a Scala project, we use **sbt**. To install **sbt**, follow the instruction in the
document *Installing sbt on Linux*.[2]

```
$ echo "deb https://dl.bintray.com/sbt/debian /" | \
> sudo tee -a /etc/apt/sources.list.d/sbt.list
$ sudo apt-key adv \
> --keyserver hkp://keyserver.ubuntu.com:80 \
> --recv 2EE0EA64E40A89B84B2DF73499E82A75642AC823
$ sudo apt update && sudo apt install -y sbt && \  ①
> cd /tmp && sbt help  ②
```

---

[2]http://www.scala-sbt.org/1.x/docs/Installing-sbt-on-Linux.html.

①   Install **sbt** with the **apt** tool.
②   Make sure the **sbt** command is available.

7.4.2   Source Directory

The Heron Scala API is available by compiling the *api-scala* target:

```
$ cd ~/heron/ && bazel build --config=ubuntu_nostyle \
> heron/api/src/scala:api-scala && \ ①
> ls -l bazel-bin/heron/api/src/scala/api-scala.jar ②
```

①   Compile the Scala API JAR file.
②   The generated JAR file path.

Create the project directory with the **sbt** command and put the dependency into the *lib* directory:

```
$ cd ~ && sbt new scala/hello-world.g8
...
name [Hello World template]: heron-streamlet-scala ①
...
$ cd ~/heron-streamlet-scala && mkdir lib
$ cp ~/heron/bazel-bin/heron/api/src/scala/api-scala.jar \
> ./lib/ ②
$ cp ~/heron/bazel-bin/heron/api/src/java/heron-api.jar \
> ./lib/ ③
```

①   Set the project name that is also the project directory name.
②   Unmanaged dependencies are JARs dropped into the *lib* directory.
③   The Scala API depends on the Java Streamlet API that should also be included in the *lib* directory.

Just as we did in Sect. 4.2.2, we need the *kryo* serialization library in the dependency. Update the *build.sbt* to include the dependency:

```
...
name := "scala-streamlet"
organization := "heronbook"
...
libraryDependencies += "com.esotericsoftware" % "kryo" % "3.0.3"
...
```

To create an uber JAR of your project with all of its dependencies, use the sbt plugin sbt-assembly.[3] Make a new file named *assembly.sbt* in the directory *project/* and add sbt-assembly as a plugin:

```
addSbtPlugin("com.eed3si9n" % "sbt-assembly" % "0.14.9")
```

---

[3]https://github.com/sbt/sbt-assembly.

After the above operations, the source directory looks like:

```
$ tree --charset=ascii ~/heron-streamlet-scala/
/home/ubuntu/heron-streamlet-scala/
|-- build.sbt
|-- lib
|   |-- api-scala.jar
|   `-- heron-api.jar
|-- project
|   |-- assembly.sbt
|   `-- build.properties
`-- src
    `-- main
        `-- scala
            `-- Main.scala
```

7.4.3 Compose Processing Graph

Similar to Sect. 7.2, we construct the processing graph in `main()` of *Main.scala*:

```scala
import java.util.concurrent.ThreadLocalRandom

import org.apache.heron.api.utils.Utils

import org.apache.heron.streamlet.Config
import org.apache.heron.streamlet.scala.{Builder, Runner}

object ExclamationTopology {
  def main(args: Array[String]): Unit = {
    val words =
      Array("nathan", "mike", "jackson", "golda", "bertels")
    val builder = Builder.newBuilder  (1)

    val s1 = builder.newSource(() => {
      Utils.sleep(500)
      words(ThreadLocalRandom.current().nextInt(words.length))
    }).setName("s1")  (2)
    val s2 = builder.newSource(() => {
      Utils.sleep(500)
      val word2 =
        words(ThreadLocalRandom.current().nextInt(words.length))
      val word3 =
        words(ThreadLocalRandom.current().nextInt(words.length))
      Array(word2, word3)
    }).setName("s2")  (3)

    val s3 = s1.map(x => x+" !!!")  (4)
    val s4 = s2.map(x => x(0)+" & "+x(1)+" !!!")  (5)
    s3.union(s4).map(x => x+" !!!").log()  (6)
```

```
val config = Config.newBuilder().setNumContainers(2)
                    .build() ⑦

new Runner().run("my-scala-streamlet", config, builder) ⑧
   }
}
```

①   The Heron Scala API wraps the Java API, and the Builder is from the Scala package.
②   The first source emitting one word in a tuple.
③   The second source emitting two words in a tuple.
④   Append "!!!" to all tuples of stream s1.
⑤   Combine two words in all tuples of stream s2 and then append "!!!".
⑥   Union two streams and append "!!!".
⑦   Configure the processing graph with two containers.
⑧   Submit the job with the name my-scala-streamlet.

7.4.4 Examine the JAR File

Compile the project, and the generated JAR file is ready to be submitted:

```
$ sbt clean assembly ①
$ ls -l target/scala-2.13/scala-streamlet-assembly-1.0.jar ②

$ heron submit local \
> target/scala-2.13/scala-streamlet-assembly-1.0.jar \
> ExclamationTopology

$ heron kill local my-scala-streamlet
```

①   Compile the source and package a JAR file with dependencies.
②   The generated JAR file path. It is ready to be submitted.

7.5 Summary

The Heron Topology API has been proven to make topologies serving large associations' requirements for a long time. We speculate that it will remain in use for many years.

We are eager to see the development of the Streamlet API. With functional programming turning out to be the trend, the Streamlet API aligns with the advancement in programming. If the Topology API works for you, stay with it. However, we encourage you to give the Streamlet API a try.

References

1. The Heron Streamlet API for Java.   https://heron.incubator.apache.org/docs/topology-development-streamlet-api. Visited on 2020-07
2. Heron streamlets. https://heron.incubator.apache.org/docs/heron-streamlet-concepts. Visited on 2020-07
3. Streamlet.java.   https://github.com/apache/incubator-heron/blob/master/heron/api/src/java/org/apache/heron/streamlet/Streamlet.java. Visited on 2020-07

Part III
Operate Heron Clusters

Chapter 8
Manage a Topology

In the last part, we wrote several topologies and compiled them into JAR files. In this chapter, we will submit topology JAR files to Heron to run the topology on the local machine. In addition, we will get familiar with the Heron command-line client and see how to manipulate a running topology. Additionally, we will learn about the topology life cycle.

8.1   Install Heron Client

In Sect. 3.6, we saw the *heron-install.sh* executable script in the directory *~/heron/ bazel-bin/scripts/packages/*. We use this script to install the Heron CLI on the local machine. Run this script to install the Heron CLI:

```
$ ~/heron/bazel-bin/scripts/packages/heron-install.sh --user
```

This command installs the Heron CLI in the current user directory *~/.heron/*. It also creates several soft links in the *~/bin/* directory pointing to the Heron CLI for easy access, which is why we need to add *~/bin/* to the environment variable PATH. To check and prepare the Heron CLI environment:

```
$ ls -l `which heron` ①
/home/ubuntu/bin/heron -> /home/ubuntu/.heron/bin/heron
$ ls ~/bin/ ②
heron
heron-admin   heron-apiserver ③
heron-tracker   heron-explorer   heron-ui ④
```

①   If the **heron** command is not found, use **export PATH="$PATH:$HOME/bin"** to declare the path. Alternatively, use **source ~/.profile**, which updates the PATH.

②   The Heron CLI includes six commands. We use the **heron** command in this chapter.

© The Author(s), under exclusive license to Springer Nature Switzerland AG 2021
H. Wu, M. Fu, *Heron Streaming*, https://doi.org/10.1007/978-3-030-60094-5_8

③   These two commands are to manage the Nomad[1] local cluster and the
      Kubernetes topologies.
④   These three commands are tools that we will see in Chap. 9.

The Heron client installation directory organization is as follows:

```
$ tree --charset=ascii ~/.heron/
/home/ubuntu/.heron/
|-- bin ①
|-- conf ②
|   |-- aurora
|   |-- examples
|   |-- kubernetes
|   |-- local ③
|   |   |-- heron_internals.yaml
|   |   |-- scheduler.yaml
|   |   |-- statemgr.yaml
|   |   `-- ...
|   |-- test
|   `-- ...
|-- dist
|   `-- heron-core.tar.gz ④
|-- examples
|   |-- heron-api-examples.jar ⑤
|   `-- ...
|-- include   ⑥
|-- lib ⑦
`-- release.yaml ⑧
```

①   Binary executable files, including **heron** and tools.
②   A *conf/* directory is organized by scheduler types. It also includes an
      examples/ directory, which contains general configuration templates, and
      a *test/* directory, which contains test configuration files.
③   A *local* scheduler is used in this book.
④   A distribution file to be uploaded to the containers. We will see what is
      inside the compression file soon in Sect. 8.1.1.
⑤   A topology JAR file, including all Heron examples composed with the
      Heron low-level API.
⑥   C++ header files for the C++ API.
⑦   Java JAR files for each Heron component, and a C++ archive file for the
      C++ API.
⑧   The Heron CLI release version, which should match the Heron API version
      to compile the topology source code.

---

[1] https://www.nomadproject.io/.

8.1.1 What Is Inside **heron-core.tar.gz**

heron-core.tar.gz contains all the executable files, including binary executable files and Java JAR files, which are used to launch processes in containers. Let us take a close look at *heron-core.tar.gz*:

```
$ tar -tf ~/.heron/dist/heron-core.tar.gz  ①
./
./release.yaml  ②
./heron-core/
./heron-core/bin/  ③
./heron-core/bin/heron-executor  ④
./heron-core/bin/...
./heron-core/lib/  ⑤
./heron-core/lib/...
```

①   The -**t** option lists the archive contents without an actual extraction.

②   The release version file, which is the same as *release.yaml* in the Heron CLI installation directory *~/.heron/* and which should also match the Heron API version we used to compile the topology source code. In simple words, the three versions—API, core, and client—should match.

③   Contains the binary executable files.

④   **heron-executor** is the entry point of a container, which is the first process to be launched in the container and is responsible for launching other processes in the container.

⑤   Contains the Java JAR files.

The container needs two primary files to run a topology properly, as shown in Fig. 2.1. One is the *heron-core.tar.gz* file, which represents the Heron layer. The other is the topology JAR file, which we compiled in the previous chapters and which represents the topology layer.

8.1.2 YAML Configuration

Each configuration item in Heron is in the format of a key-value pair. According to the key prefix, the configurations fall into two categories: topology and heron. Configuration items with the key starting with topology, for example, topology.container.cpu, are used by topology writers composing topology code. Configuration items with the key starting with heron, for example, heron.logging.directory, are set in the YAML files we will talk about in this section. To run a topology properly, both categories of configurations are required.

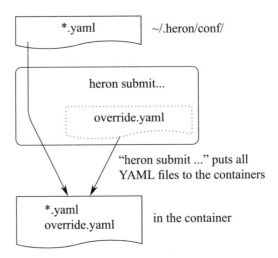

Fig. 8.1 Heron YAML configurations

YAML configurations can be set at two levels [1]:

System level
    System-level configurations apply to the whole Heron cluster rather than to any specific component (e.g., logging configurations). All the Heron components share them.
Component level
    Component-level configurations enable you to establish default configurations for different components.

System-level configurations are mainly in *heron_internals.yaml*, while component-level configurations are scattered in all YAML files. Most Heron components read *heron_internals.yaml*, while each Heron component may read some specific YAML files.

Once the topology is deployed, most system-level and component-level configurations are fixed at any stage of the topology's life cycle. However, you have a chance to override the configuration by supplying the `--config-property` option when you submit a topology using the `heron submit` command. The `--config-property` option generates a YAML file called *override.yaml* and injects this YAML file into the containers, as shown in Fig. 8.1. *override.yaml* does not appear in the Heron CLI installation directory *~/.heron/conf/*.

The cluster configuration includes all configuration files. The following list shows the standard YAML files [2]:

```
$ tree --charset=ascii ~/.heron/conf/local/
/home/ubuntu/.heron/conf/local/
|-- client.yaml   ①
|-- downloader.yaml  ②
|-- healthmgr.yaml  ③
```

```
|-- heron_internals.yaml  (4)
|-- metrics_sinks.yaml  (5)
|-- packing.yaml  (6)
|-- scheduler.yaml  (7)
|-- stateful.yaml  (8)
|-- statemgr.yaml  (9)
`-- uploader.yaml  (10)
```

(1)   This file controls the behavior of the Heron client. This is optional.

(2)   This file maps the file transfer protocols to implementation class names.

(3)   This file contains the parameters for Heron Health Manager, including the metrics source and policy setting.

(4)   This file contains the parameters for the Heron kernel. Tuning these parameters requires lots of experiments and advanced knowledge of Heron. For beginners, the best option is just to copy the file provided with the sample configuration.

(5)   This file specifies where the runtime system and topology metrics will be routed. By default, `file-sink` and `tmaster-sink` need to be present. Also, `scribe-sink` and `graphite-sink` are supported.

(6)   This file specifies the classes for the packing algorithm, which defaults to round-robin, if not specified.

(7)   This file specifies the configurations for the launcher and scheduler. Any parameters for a particular scheduler should be specified in this file.

(8)   This file contains the parameters for Heron Checkpoint Manager and stateful processing state storage, which is used by the *effectively-once* semantic (more in Sect. 6.1.2).

(9)   This file contains the configurations for State Manager, which maintains the Topology Master location, logical/packing/physical plans, and scheduler state.

(10) This file specifies the configurations for the uploader, which is responsible for pushing the topology JARs to storage. The containers will download these JARs from storage at runtime.

Let us take a closer look at some typical configuration key-value pairs in the *~/.heron/conf/local/* YAML files:

```
$ grep -F 'package.core' client.yaml
heron.package.core.directory: ${HERON_DIST}/heron-core  (1)

$ grep -F 'logging.directory' heron_internals.yaml
heron.logging.directory: "log-files"  (2)

$ grep -F 'local.working.directory' scheduler.yaml | fmt
heron.scheduler.local.working.directory:  (3)
${HOME}/.herondata/topologies/${CLUSTER}/${ROLE}/${TOPOLOGY}

$ grep -F 'root.path' statemgr.yaml | fmt
heron.statemgr.root.path:  (4)
${HOME}/.herondata/repository/state/${CLUSTER}
```

```
$ grep -F 'localfs.file.system.directory' uploader.yaml | fmt
heron.uploader.localfs.file.system.directory: ⑤
.../repository/topologies/${CLUSTER}/${ROLE}/${TOPOLOGY}
```

①   Location of the core package, which is injected into the container or the working directory for local mode.

②   The relative path to the logging directory in the container.

③   The working directory for topologies. The local scheduler runs processes using this directory as the current directory.

④   Path of the root address to store state.

⑤   Name of the directory to upload the user topology JAR file.

The configuration value may include variables that are substituted by Heron components when they are executed. We can track the route of the *heron-core.tar. gz* tar file and the user topology JAR file from these configurations. The *heron-core. tar.gz* file is downloaded from `heron.package.core.uri` to `heron.scheduler. local.working.directory`. The user topology JAR file is first uploaded to `heron.uploader.localfs.file.system.directory` and then downloaded to `heron.scheduler.local.working.directory`.

8.2   Run Topology

To run a topology, **heron submit** is the proper command. Before trying **heron submit**, let us try the **heron** command to display its help message:

```
$ heron
usage: heron [-h] <command> <options> ...

optional arguments:
  -h, --help              show this help message and exit

Available commands:
    activate            Activate a topology
    config              Config properties for a cluster
    deactivate          Deactivate a topology
    help                Prints help for commands
    kill                Kill a topology
    restart             Restart a topology
    submit              Submit a topology
    update              Update a topology
    version             Print version of heron-cli

Getting more help:
    heron help <command> Prints help and options for <command>
```

NOTE

If the command complains that JAVA_HOME is not set, use the following command to set it:

```
$ export JAVA_HOME=/usr/lib/jvm/java-11-openjdk-amd64
```

8.2.1 Topology Life Cycle

You can use the **heron** command-line tool to manage the entire life cycle of a topology, which includes the following stages, as also shown in Fig. 8.2:

- Submit a topology. The topology starts processing data once submitted since it is by default deployed as activated.
- Activate a topology. If the topology is deactivated, it will begin processing data.
- Restart a topology. The topology will be in the same state as before restarting.
- Deactivate a topology. If the topology is activated, it will stop processing data but remain in the cluster without freeing any resources.
- Kill a topology. All topology resources are freed.

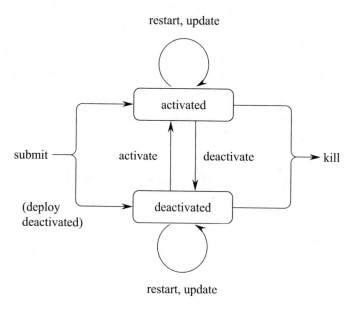

Fig. 8.2 Topology life cycle

8.2.2   Submit Topology

Now let us follow the above topology life cycle to submit an example topology that we wrote in Sect. 5.2.1:

```
$ heron help submit
usage: heron submit [options] cluster/[role]/[env]
  topology-file-name topology-class-name [topology-args]

$ heron submit local ~/heron-topology-java/target\
> /heron-topology-java-1.0-SNAPSHOT-jar-with-dependencies.jar \
> heronbook.ExclamationTopology my-test-topology-1
```

The following are the arguments of the `submit` command [3]:

cluster/[role]/[env]
> The cluster where the topology needs to be submitted, optionally taking the role and the environment. For example, *local/ads/PROD* or just *local*. In our experiment, we use the local mode.

topology-file-name
> The path of the file in which you have packaged the topology's code. For Java and Scala topologies, this will be a JAR file; for Python topologies, this will be a PEX file; for topologies in other languages, this could be a TAR file. For example, */path/to/topology/my-topology.jar*.

topology-class-name
> The name of the class containing the main function for the topology. For example, *com.example.topologies.MyTopology*.

topology-args (optional)
> Arguments specific to the topology. You will need to supply additional args only if the main function for your topology requires them. Usually, a topology name is supplied, such as *my-test-topology-1* in our experiment. For *ExclamationTopology*, if you do not supply a topology name, the **heron** command will run this topology in the simulator. The simulator mode runs the topology in a single process for easy debug.

NOTE

If the command complains that the hostname cannot be resolved (often seen on a Macintosh), add it to */etc/hosts*:

```
$ sudo cp /etc/hosts /etc/hosts.bak && \
> echo "127.0.0.1 `hostname`" | sudo tee --append /etc/hosts
```

8.2.3   Observe the Topology Running Status

Check if the topology is running with the `pstree` command; you will see a similar output, as follows:

```
$ pstree -A
...
|-java-+-python3-+-heron-tmaster  ①
|      |          |-java---22*[{java}]
|      |          |-java---23*[{java}]
|      |          |-java---26*[{java}]
|      |          |-python3
|      |          `-6*[{python3}]
|      |-3*[python3-+-heron-stmgr]  ②
|      |          |-java---27*[{java}]]
|      |          |-2*[java---19*[{java}]]]
|      |          |-python3]
|      |          `-6*[{python3}]]
|      `-27*[{java}]
...
```

①   Container 0 includes `heron-tmaster`.
②   Containers 1–3 include `heron-stmgr`.

If you want to know more about the Heron processes, use `pstree -Aa`, which shows the parameters to launch each Heron process.

8.3   Explore the "heron" Command

`heron submit` has some useful options. Besides the `heron submit` command, there are several other commands, such as `heron kill` and `heron update`.

8.3.1   Common Arguments and Optional Flags

The topology life cycle management subcommands, including `submit`, `activate`, `deactivate`, `restart`, `update`, and `kill`, share the following required arguments [3]:

cluster
    The name of the cluster where the command is executed.
role
    User or group, depending on deployment. If not provided, it defaults to the current user, which is the same one as the command `whoami` prints.

env
> The tag for including additional information, for example, a topology can be tagged as *PROD* or *DEVEL* to indicate whether it is in production or development. If *env* is not provided, it is given a value of *default*.

These three arguments together in the format of *cluster/role/env* form a "job key" inspired by Apache Aurora.[2] For example, *local/myserviceaccount/prod* refers to the job launched by the role *myserviceaccount* in the cluster *local* running in the environment *prod*. If the job does not have a role or environment, you can just specify the cluster, and the argument is merely *local*.

In addition to the standard arguments, **heron** supports a common set of optional flags for all subcommands [3]:

--config-path
> Every Heron cluster must provide a few configuration files that are kept under a directory named after the cluster. By default, when a cluster is provided in the command, it searches the *conf/* directory for a directory with the cluster name. This flag enables you to specify a nonstandard directory to search for the cluster directory. The default is *~/.heron/conf*.

--config-property
> Heron supports overriding several configuration parameters. These parameters are specified in the format *key=value*. This flag can be supplied multiple times. The default is an empty key-value pair.

--verbose
> When this flag is provided, **heron** prints logs that provide detailed information about the execution.

--service-url
> API service endpoint. This option implies an API server running in the data center, which is responsible for executing the delegated actions from the user command.

8.3.2   Explore "heron submit" Options

The first subcommand you need for a new topology is **submit**, which launches the topology in a Heron cluster. Topologies are submitted in the activated state by default; however, you can also specify the initial state to be the deactivated state (more in Sect. 8.3.4). Several useful optional options for **heron submit** are as follows:

--deploy-deactivated
> Deploy topology in the deactivated mode (the default is "false").

---

[2]http://aurora.apache.org/documentation/latest/reference/client-commands/#job-keys.

--topology-main-jvm-property

The Heron CLI invokes the topology main(), and you can pass a system property to its java -D.

--dry-run

Enable dry-run mode, which prints the packing plan without actually running it.

The following is an example that submits a topology to the *devcluster* cluster (no role or environment specified). The "submit" subcommand will search the path *~/heronclusters* and adopt the configuration for the cluster *devcluster*. The JAR path and its entry class are specified along with arguments to main().

```
$ heron submit \
> --config-path ~/heronclusters \
> devcluster \
> /path/to/topology/my-topology.jar \
> com.example.topologies.MyTopology \
> my-topology
```

--dry-run is encouraged before you submit a topology. It helps you to get an idea about the number of resources that will be allocated to each container. The following is an example output from the **--dry-run** option:

```
$ heron submit local --dry-run ~/heron-topology-java/target\
> /heron-topology-java-1.0-SNAPSHOT-jar-with-dependencies.jar \
> heronbook.ExclamationTopology my-test-topology-2
Total number of containers: 3  ①
Using packing class: org.apache.heron...RoundRobinPacking  ②
Container 1  ③
CPU: 3.0, RAM: 4096 MB, Disk: 14336 MB  ④
========================================================  ⑤
| component | task ID | CPU | RAM (MB) | disk (MB) |
--------------------------------------------------------
|  exclaim2 |       1 | 1.0 |     1024 |      1024 |
|  exclaim1 |       4 | 1.0 |     1024 |      1024 |
========================================================

Container 2  ⑥
CPU: 3.0, RAM: 4096 MB, Disk: 14336 MB
========================================================
| component | task ID | CPU | RAM (MB) | disk (MB) |
--------------------------------------------------------
...
```

①   The number of containers in this topology.
②   The packing algorithm used by this topology.
③   Beginning of container 1 resource allocation.
④   Total resources allocated to container 1.
⑤   The resources allocated for tasks in container 1.
⑥   Beginning of container 2 resource allocation.

8.3.3 Kill Topology

Once you have submitted a topology successfully and would like to remove it entirely, you can use the **kill** subcommand by providing the topology name to the command:

```
$ heron kill local my-test-topology-1
```

The Heron topology processes should not be present in the pstree output.

8.3.4 Activate and Deactivate Topology

This pair of commands toggle the topology state between **activated** and **deactivated**. The activated state is the default setting during the submission process. You can deactivate a running topology at any time using the **deactivate** command. Both commands need a topology name parameter, for example:

```
$ heron deactivate local my-test-topology-1
$ heron activate local my-test-topology-1
```

8.3.5 Restart Topology

heron restart can either restart the whole topology or restart a container. If you restart a topology in the deactivated state, the topology remains deactivated after restarting. The **restart** command kills containers and allocates new containers, while a rapid **deactivate** and then **activate** neither kills nor allocates containers.

```
$ heron help restart
usage: heron restart [options] cluster/[role]/[env]
  <topology-name> [container-id]

$ heron restart local my-test-topology-1 2 ①
```

①   Restart container 2. In the pstree output, you can find the PID changes for the container 2 processes. The assigned ports for container 2 are also changed.

Use the container-id (optional) argument to specify which container you want to restart. If this argument is not supplied, all containers will be restarted.

8.3.6 Update Topology

At runtime, you can update either the parallelism in the packing plan or the topology/component key-value configurations. To update the packing plan, you can specify either the component parallelism or the container count; then, the packing algorithm will calculate other values to complete a new packing plan. The `update` command is usually implemented in three steps: try to `deactivate` the topology, recalculate the packing plan, and `activate` the topology.

```
$ heron help update
usage: heron update [options] cluster/[role]/[env]
  <topology-name>
    [--component-parallelism <name:value>]
    [--container-number value]
    [--runtime-config [component:]<name:value>]

$ heron update local my-test-topology-1 \
> --component-parallelism=word:1 \
> --component-parallelism=exclaim1:3
```

You can repeat the `--component-parallelism` argument multiple times, each time for a particular component.

8.4 Summary

In this chapter, we ran the topology JAR file built in the last part. Furthermore, we studied the topology life cycle and how to manipulate it through the `heron` command as well as its arguments and options. In the next chapter, we will explore the tools to manage multiple topologies typically used in data centers.

References

1. Cluster config overview. https://heron.incubator.apache.org/docs/cluster-config-overview. Visited on 2020-07
2. Configuring a cluster. https://heron.incubator.apache.org/docs/deployment-configuration. Visited on 2020-07
3. Managing topologies with Heron CLI. https://heron.incubator.apache.org/docs/user-manuals-heron-cli. Visited on 2020-07

Chapter 9
Manage Multiple Topologies

In the previous chapters, we studied almost every aspect of a topology. However, we worked on only one topology. Heron is a scalable system and can run multiple topologies. In a real data center, there could be hundreds of topologies. In this chapter, we will run multiple topologies in local mode and try Heron tools to manage them.

Heron tools mainly include Heron Tracker, Heron UI, and Heron Explorer. Heron Tracker tracks State Manager and exposes a query endpoint. Heron UI and Heron Explorer connect to Heron Tracker and display query results. If the topologies share the same State Manager, they can be displayed by Heron tools, as shown in Fig. 9.1. All three tools are read-only tools; in other words, they do not change the topology state. Meanwhile, State Manager organizes states of multiple topologies by directories; thus, their states are isolated.

We may submit the same topology JAR file multiple times with different topology names to create a multiple-topology environment. The following commands run the same topology code with the names: *my-test-topology-3* and *my-test-topology-4*:

```
$ heron submit local ~/heron-topology-java/target\
> /heron-topology-java-1.0-SNAPSHOT-jar-with-dependencies.jar \
> heronbook.ExclamationTopology my-test-topology-3

$ heron submit local ~/heron-topology-java/target\
> /heron-topology-java-1.0-SNAPSHOT-jar-with-dependencies.jar \
> heronbook.ExclamationTopology my-test-topology-4
```

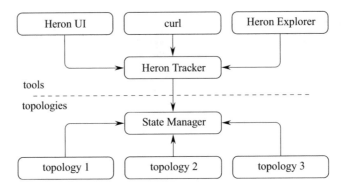

Fig. 9.1 Heron Tracker-based tools

9.1 Install Heron Tools

In Sect. 8.1, we saw the *heron-install.sh* executable script install Heron tools to
~/.heron/. Let us revisit the Heron installation directory and highlight the following
Heron tools:

```
$ tree --charset=ascii -L 2 -- ~/.heron/
/home/ubuntu/.heron/
|-- bin ①
|   |-- heron-explorer
|   |-- heron-tracker
|   `-- heron-ui
|-- conf
|   |-- heron_tracker.yaml ②
|   `-- local
...
```

①   Binary executable files, including **heron-tracker**, **heron-ui**, and **heron-explorer**.
②   Configuration for **heron-tracker**.

9.2 Heron Tracker

Before we run Heron Tracker, let us read the help usage first:

```
$ heron-tracker help
usage: heron-tracker [options] [help]

Optional arguments:
  --config-file (a string; path to config file; default:
    "/home/ubuntu/.heron/conf/heron_tracker.yaml") ①
  ...
```

①   The default configuration for Heron Tracker, which specifies information
on the local file State Manager, including the root path.

Thanks to the default *heron_tracker.yaml* configuration, we do not have to supply
parameters to the **heron-tracker** command:

```
$ heron-tracker &
...
[INFO]: Running on port: 8888  ①
...
```

①   The Heron Tracker server listens on the default 8888 port.

Heron Tracker exposes the JSON RESTful endpoint. We demonstrate how to
interact with the Heron Tracker endpoint with two examples. The first example lists
all the topologies:

```
$ curl http://127.0.0.1:8888/topologies | \
> python3 -m json.tool --sort-keys
{
    "executiontime": 5.602836608886719e-05,
    "message": "",  ①
    "result": {
        "local": {  ②
            "ubuntu": {  ③
                "default": [  ④
                    "my-test-topology-4",  ⑤
                    "my-test-topology-3"
                ]
            }
        }
    },
    "status": "success",
    "tracker_version": "master"  ⑥
}
```

①   The message returned by the endpoints. It is usually empty; however,
failures always have a message.
②   Cluster name.
③   Role name.
④   The *environ* string.
⑤   A list of topology names.
⑥   The Tracker API version.

The following example shows how to query metrics:

```
$ curl -X GET \  ①
> -d "cluster=local" \
> -d "environ=default" \
> -d "topology=my-test-topology-4" \
> -d "starttime=`date --date='1 hour ago' +%s`" \  ②
> -d "endtime=`date +%s`" \  ③
```

```
> -d "query=TS(__stmgr__,*,\
>   __server/__time_spent_back_pressure_initiated)" \ ④
> http://127.0.0.1:8888/topologies/metricsquery | \
> python3 -m json.tool --sort-keys
{
    ...
    "result": {
        "endtime": 1597004253,
        "starttime": 1597000653,
        "timeline": [ ⑤
            {
                "data": {
                    "1597003320": 0.0, ⑥
                    "1597003380": 0.0,
                    ...
                },
                "instance": "stmgr-1" ⑦
            },
            ...
        ]
    },
    ...
}
```

①   Heron Tracker accepts HTTP GET only.
②   The start time was 1 h ago.
③   The end time is now.
④   The query string is in Metrics Query Language.
⑤   The query result is a time series of type "multivariate time series."
⑥   The "timestamp-value" pair.
⑦   Metric reporter or metric source.

When Heron Tracker receives metrics queries, it relays the requests to Topology Master, which responds with the aggregated metrics. Topology Master provides some limited metrics aggregation capability. If you need advanced metrics monitoring and alerts, you can send the metrics to an external metrics sink (more in Chap. 11). The query results are time series, and they are categorized into two types [1]:

Univariate time series
     A time series is called univariate if there is only one set of data. For example, a time series representing the sums of several time series would be a univariate time series.
Multivariate time series
     A set of multiple time series is collectively called multivariate. Note that these time series are associated with their instances.

The Metrics Query Language supports many operators. Our example shows the basic TS (Time Series) operator, which accepts a list of elements [1]:

- Component name
- Instance—Can be "*" for all instances (multivariate time series), or a single instance ID (univariate time series)
- Metric name—Full metric name with stream ID if applicable

Details about the Metrics Query Language can be found in the Heron documentation.[1]

To kill the Heron Tracker server, run:

```
$ kill $(pgrep -f heron-tracker)
```

9.3   Heron UI

Heron UI does not accept YAML configuration. All customized configurations are supposed to supply in the command line. Before we run Heron UI, let us read the help usage first:

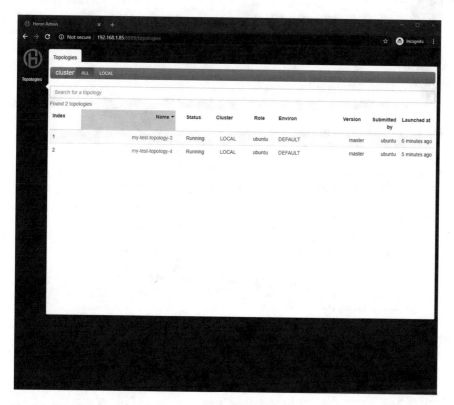

Fig. 9.2 Heron UI landing page

---

[1] https://heron.incubator.apache.org/docs/user-manuals-tracker-rest.

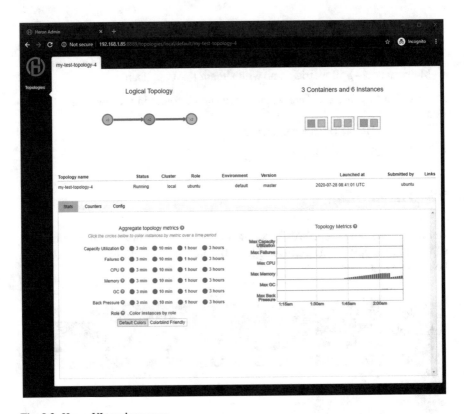

Fig. 9.3 Heron UI topology page

```
$ heron-ui help
usage: heron-ui [options] [help]

Optional arguments:
  --tracker_url (a url; path to tracker; default:
    "http://127.0.0.1:8888") ①
  --address (a string; address to listen; default:
    "0.0.0.0") ②
  --port (an integer; port to listen; default: 8889) ③
  --base_url (a string; the base url path if operating behind
    proxy; default: ) ④
```

①   Heron UI depends on Heron Tracker. The default value matches the default endpoint of Heron Tracker.

②   Heron UI binds to address.

③   Heron UI binds to the default port 8889.

④   Heron UI URL path to bypass proxy.

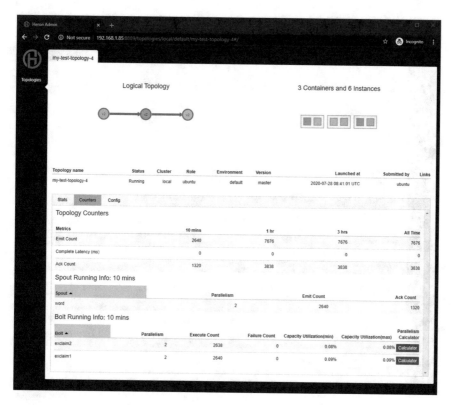

Fig. 9.4 Heron UI topology page: "Counters" tab

Launch Heron UI by:

```
$ heron-ui &
[INFO]: Listening at http://0.0.0.0:8889
[INFO]: Using tracker url: http://127.0.0.1:8888
```

Open a browser and point to the Ubuntu Server address and port 8889. For example, on the author's laptop, http://192.168.1.85:8889/topologies, as shown in Fig. 9.2. Click the topology name and enter the topology page, as shown in Fig. 9.3. In the upper part of the topology page, there are two figures side by side. The figure on the left shows the logical plan and the figure on the right shows the packing plan. In the bottom half of the page, there are three tabs: Stats, Counters, and Config. The "Stats" tab displays the topology running stats, including failures/exceptions, CPU, memory, JVM GC, and backpressure. The "Counters" tab displays the performance counters of the whole topology and each component, including emit-count, ack-count, and latency, as shown in Fig. 9.4. The "Config" tab displays all the topology configuration key-value pairs, as shown in Fig. 9.5.

Click a task in the packing plan, and the web page highlights the information for this task; for example, *container_1_exclaim1_4* is shown in Fig. 9.6. When you

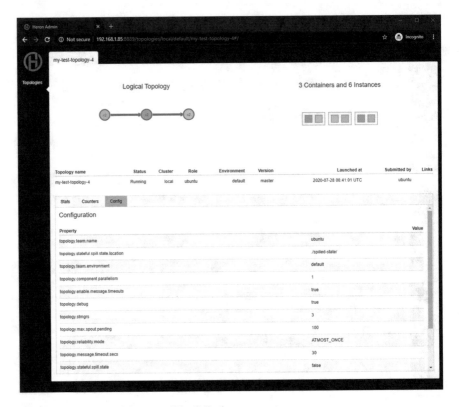

Fig. 9.5 Heron UI topology page: "Config" tab

scroll down to the bottom, you can find the metrics and drop-down quick links
shown in Fig. 9.7. The "Job" quick link directs the browser to the container working
directory, where all the container files and logs exist, as shown in Fig. 9.8.

To kill the Heron UI server, run:

```
$ kill $(pgrep -f heron-ui)
```

Note that Heron UI depends on Heron Tracker. If you want to stop all Heron tools,
you need to kill Heron Tracker as well.

9.4   Heron Explorer

If you are comfortable with the command-line environment, you may want to use
Heron Explorer, which provides the same features as Heron UI. Heron Explorer is
installed along with the Heron CLI. The help usage is self-explanatory, which is a
good starting point:

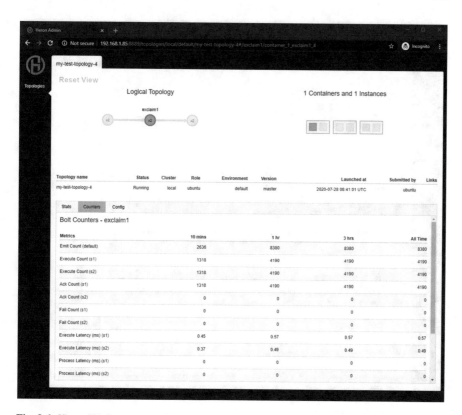

Fig. 9.6   Heron UI: focus on a task

```
$ heron-explorer help
usage: heron-explorer <command> <options> ...

Available commands:
    clusters            Display existing clusters
    components          Display information of a topology's
                        components
    metrics             Display info of a topology's metrics
    containers          Display info of a topology's containers
                        metrics
    topologies          Display running topologies
    help                Display help
    version             Display version
```

The default configuration is compatible with the local default Heron Tracker. If you
want to change the default configuration, you have to pass the new configuration
along with the command. Most commands accept cluster and topology name as
arguments. We show several examples:

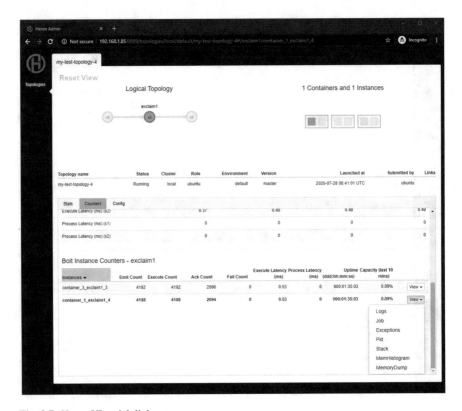

Fig. 9.7 Heron UI: quick links

```
$ heron-explorer topologies local ①
[INFO]: Using tracker URL: http://127.0.0.1:8888
Topologies running in cluster 'local'
role    env        topology
------  -------    ------------------
ubuntu  default    my-test-topology-4
ubuntu  default    my-test-topology-3

$ heron-explorer components local my-test-topology-4 ②
[INFO]: Using tracker URL: http://127.0.0.1:8888
type    name       parallelism  input      output
------  -------    -----------  ---------  -----------------
spout   word           2        -          exclaim1,exclaim1
bolt    exclaim2       2        exclaim1   -
bolt    exclaim1       2        word,word  exclaim2
```

① Supply cluster.
② Supply cluster and topology names.

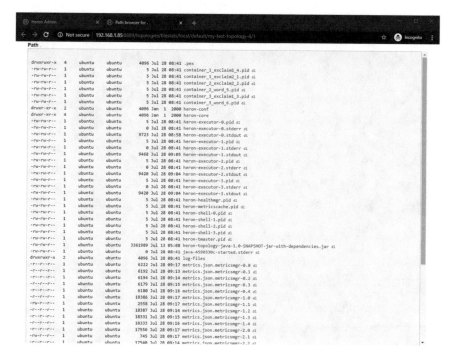

Fig. 9.8 Heron UI job page

9.5 Summary

In this chapter, we ran multiple topologies. We installed Heron tools and studied the dependencies between tools. Then we tried three tools one by one. For Heron Tracker, we tried its JSON RESTful endpoint. For Heron UI, we saw the topology page and tried quick links. For Heron Explorer, we tried several examples.

After running the topology, you may be curious about how the topology runs, how the tuples are transmitted in the topology DAG, and how to investigate the tuple flows in the topology code from the topology writer's perspective. The next part will clear all your doubts.

Reference

1. Heron Tracker REST API. https://heron.incubator.apache.org/docs/user-manuals-tracker-rest. Visited on 2020-07

Part IV
Heron Insights

Chapter 10
Explore Heron

In the previous chapters, we studied how to write a topology and how to operate a running topology. In this chapter, we move from a topology writer's and operator's perspectives to a Heron developer's perspective and study the core Heron components.

We will first study how Heron manages the system state with State Manager in Sect. 10.2 and how Heron schedules topologies by Heron Scheduler in Sect. 10.3, both of which are quite independent of other Heron components. Then, we will study how Heron manages a topology and how the tuple flow is routed by the Topology Master, Stream Manager, and Heron Instance in Sect. 10.4. In the last section, Sect. 10.5, we will study the metrics system, including Metrics Manager and MetricsCache Manager.

10.1 Heron Processes

The chapter's introductory paragraphs listed the core Heron components which we will discuss in this chapter. However, you may be curious about how many components there are in Heron exactly. Here, we provide a simple way to get a bird's-eye view of Heron: find the process entry point in the source code. As we have seen in Sect. 3.1, the major coding languages in Heron are Java, C++, and Python. Run the following searching commands in the source code root directory ~/heron/ to find the process entry points for each type of coding language.

10.1.1 Java Processes

```
$ grep -r --include=*.java \
> "main\s*(\s*String\s*\[\]" heron/ | \ ①
> sed 's@:.*$@@g' | sed 's@^heron@...@g' | \
> sed 's@src.*heron@...@g' | sort
.../ckptmgr/.../ckptmgr/CheckpointManager.java ②
.../downloaders/.../downloader/DownloadRunner.java ③
.../healthmgr/.../healthmgr/HealthManager.java ④
.../instance/.../instance/HeronInstance.java ⑤
.../instance/.../instance/util/JvmVersion.java ⑥
.../io/dlog/.../dlog/Util.java ⑦
.../metricscachemgr/.../MetricsCacheManagerHttpServer.java ⑧
.../metricscachemgr/.../MetricsCacheManager.java ⑨
.../metricsmgr/.../metricsmgr/MetricsManager.java ⑩
.../scheduler-core/.../scheduler/RuntimeManagerMain.java ⑪
.../scheduler-core/.../scheduler/SchedulerMain.java ⑫
.../scheduler-core/.../scheduler/SubmitterMain.java ⑬
.../statemgrs/.../localfs/LocalFileSystemStateManager.java ⑭
.../statemgrs/.../zookeeper/.../CuratorStateManager.java ⑮
.../tools/apiserver/.../apiserver/Runtime.java ⑯
```

①   Search the entry function `public static void main(String [] args)` in the Java files.

②   The Checkpoint Manager process in container 0, which is responsible for managing states for stateful topologies.

③   The **heron-downloader** executable file, to download files from remote storage to the local machine, either in the Heron CLI or inside containers.

④   The Health Manager process in container 0, which is responsible for topology self-tuning, self-stabilizing, and self-healing.

⑤   The Heron Instance processes in the container executing spout and bolt tasks.

⑥   The helper process, used by Heron Executor to determine the system JVM version.

⑦   A utility program to copy data between filesystem files and dlog streams.

⑧   The helper process, for local mode debugging, to query MetricsCache Manager and print result metrics.

⑨   The MetricsCache Manager process in container 0, which is to substitute metrics collector in Topology Master.

⑩   The Metrics Manager process in every container responsible for collecting metrics inside the container and pushing the metrics to MetricsCache Manager.

⑪   The client process, invoked by the Heron CLI, to operate a running topology.

⑫   The cluster or client process scheduler, responsible for scheduling a topology.

⑬   The client process, invoked by the Heron CLI, to submit a new topology.

⑭   The helper process for debugging in local mode. It prints the states in State Manager.

⑮ The helper process for debugging with ZooKeeper. It prints the states in State Manager.

⑯ The cluster server, working with Kubernetes scheduler.

10.1.2 C++ Processes

```
$ grep -r --include=*.cpp "main\s*(.*,.*)" heron/ | \ ①
> grep -v '/tests/cpp' | \
> sed 's@:.*$@@g' | sed 's@src/cpp@...@g' | \
> sed 's@^heron@...@g' | sort
.../common/... ②
.../instance/.../instance-main.cpp ③
.../stmgr/.../server/stmgr-main.cpp ④
.../tmaster/.../server/tmaster-main.cpp ⑤
```

① Search the entry function main(int argc, char** argv) in the *cpp* files.
② Helper, example, and test processes.
③ The C++ implementation of the Heron Instance process.
④ The Stream Manager process in the containers except container 0.
⑤ The Topology Master process in container 0.

Another way to discover the C++ processes is to search the Bazel BUILD files to find the binary package output. The search result shows only three C++ processes:

```
$ bazel query 'kind(cc_binary, heron/...)' ①
//heron/tmaster/src/cpp:heron-tmaster
//heron/stmgr/src/cpp:heron-stmgr
//heron/instance/src/cpp:heron-cpp-instance
```

① Search the cc_binary rule in the Bazel rules under the directory *heron/*.

10.1.3 Python Processes

```
$ bazel query 'kind(pex_binary, heron/...)' | \
> grep -v '/tests/' | sort
//heron/executor/src/python:heron-executor ①
//heron/instance/src/python:heron-python-instance ②
//heron/shell/src/python:heron-shell ③
//heron/tools/... ④
```

① The Heron Executor process is the first process launched in the container. It is the entry point for the container, and it starts and monitors all the other processes in the container.
② The Heron Instance process responsible for executing Python topology spout and bolt tasks.
③ A simple web server process in containers, mainly responsible for diagnosis queries.

④   Most Heron tools are Python processes. We already saw the Heron CLI
        command **heron** in Chap. 8 and Heron tools in Chap. 9.

Although Python does not define the process entry point, the main module
often starts with if __name__ == '__main__':, which we can use as search
keywords. The searching result matches the above Bazel search result:

```
$ grep -r --include=*.py "__main__" heron/ | \
> grep -v '#.*__main__' | sed 's/:.*$//g' | sort
heron/executor/src/python/heron_executor.py
heron/instance/src/python/instance/st_heron_instance.py
heron/shell/src/python/main.py
heron/statemgrs/src/python/configloader.py  ①
heron/tools/...
```

①   The helper main method used to verify configuration files, intended for
        manual verification only.

10.2 State Manager

Heron supports two State Managers: the local State Manager and the ZooKeeper
State Manager. Both State Managers store states in a directory tree. As we saw
in Sect. 8.1.2, the local State Manager stores the states in the local directory
~/.herondata/repository/state/local/.

State Manager has two functions in Heron. The first function is for service
discovery of "three locations," including the Topology Master location, the Heron
Scheduler location, and the MetricsCache Manager location. When the containers
start, they have to find out the shared service locations, such as Topology Master.

The second function is to cache the system metadata of "three plans": logical
plan, packing plan, and physical plan. More details on the three plans are in
Sect. 10.3.1. From a system design perspective, since Topology Master maintains
the three plans, the containers may ask Topology Master for the three plans once
the Topology Master service location is discovered. However, Heron stores a copy
of the three plans in State Manager for easy access by containers rather than asking
Topology Master in container 0.

There are nine subdirectories in the root state directory. We can list the state
directories after the target topology is launched:

```
$ tree --charset=ascii -L 1 ~/.herondata/repository/state/local/
/home/ubuntu/.herondata/repository/state/local/
|-- executionstate  ①
|-- locks  ②
|-- metricscaches  ③
|-- packingplans  ④
|-- pplans  ⑤
|-- schedulers  ⑥
|-- statefulcheckpoints  ⑦
|-- tmasters  ⑧
`-- topologies  ⑨
```

①   Execution state.
②   Distributed locks. Relic directory.
③   The MetricsCache Manager location, one of the three locations.
④   Packing plan, one of the three plans.
⑤   Physical plan, one of the three plans.
⑥   The Heron Scheduler location, one of the three locations.
⑦   Stateful checkpoints, for stateful topologies.
⑧   The Topology Master location, one of the three locations.
⑨   Logical plan, one of the three plans.

State Manager provides a helper tool to print out the details of the topology state for two State Manager types. For the topology *my-java-topology* we ran in Chap. 5, we can either use Bazel to run the tool from the source code root directory or use Java to run from a JAR file, and we will see similar usage in Sect. 10.5.2. Here is an excerpt of output by Bazel:

```
$ bazel run --config=ubuntu_nostyle \
> heron/statemgrs/src/java:localfs-statemgr-unshaded \
> my-java-topology  ①
==> State Manager root path: ~/.herondata/...  ②
==> Topology my-java-topology found
==> Topology:  ③
==> ExecutionState:  ④
topology_name: "my-java-topology"
topology_id: "my-java-topologycaff798c-8c51-483d-..."  ⑤
submission_time: 1554548529
submission_user: "ubuntu"
cluster: "local"  ⑥
environ: "default"
role: "ubuntu"
release_state {  ⑦
...
==> SchedulerLocation:  ⑧
topology_name: "my-java-topology"
http_endpoint: "vm:35237"
==> TMasterLocation:  ⑨
topology_name: "my-java-topology"
topology_id: "my-java-topologycaff798c-8c51-483d-..."
host: "vm"
controller_port: 36049
master_port: 45711
stats_port: 34337
==> MetricsCacheLocation:  ⑩
topology_name: "my-java-topology"
topology_id: "my-java-topologycaff798c-8c51-483d-..."
host: "vm"
controller_port: -1
master_port: 45335
stats_port: 44599
==> PackingPlan:  ⑪
==> PhysicalPlan:  ⑫
```

① Run the Bazel rule with topology name supplied.
② State root path, which is the same as the default value we saw in the YAML file.
③ The *topologies* directory for the logical plan; we will examine it in the next section.
④ The *executionstate* directory.
⑤ A topology JAR file may be submitted and killed with the same name several times. To differentiate them, each topology has a unique ID.
⑥ For local mode, the cluster value is *local*; the default role is the same with submission user, and the default environ string is *default*.
⑦ Heron release version.
⑧ The *schedulers* directory, one of the three locations.
⑨ The *tmasters* directory, one of the three locations.
⑩ The *metricscaches* directory, which shares the same protocol buffers structure with the *tmasters* directory, one of the three locations. The -1 port is because MetricsCache Manager does not need `controller_port`, which is used for Topology Master to receive a request from the CLI when MetricsCache Manager does not respond to the CLI.
⑪ The *packingplans* directory for the packing plan; we will examine it in the next section.
⑫ The *pplans* directory for the physical plan; we will examine it in the next section.

10.3   Heron Scheduler

Although we only use the local scheduler or local mode to illustrate the Heron system in this book, Heron supports several schedulers, including:

- Kubernetes
- Apache Hadoop YARN
- Apache Aurora, Apache Marathon, Apache Mesos
- Slurm
- HashiCorp Nomad

We will see how to extend Heron with a new scheduler in Chap. 12.

10.3.1   Three Plans: Logical, Packing, and Physical

The Heron CLI runs a topology JAR file to get a logical plan (*topology.defn* file). The packing algorithm is applied to the logical plan to generate a packing plan. Heron Scheduler requires a packing plan to schedule containers. After the containers are launched, Stream Managers register with Topology Master so that a physical plan is generated. Heron stores a copy of the three plans in State Manager. Here is the complementary part of the previous State Manager printout excerpt:

```
==> Topology:
id: "my-java-topologycaff798c-8c51-483d-9336-815ad55d06b7"
name: "my-java-topology"
spouts {  ①
  comp {  ②
    ...
  }
  outputs {  ③
    stream {  ④
      ...
    }
    schema {  ⑤
      ...
    }
  }
  ...
}
bolts {
  comp {
    ...
  }
  inputs {  ⑥
    stream {
      ...
    }
    gtype: SHUFFLE
  }
  outputs {
    ...
  }
}
bolts {
  ...
}
state: RUNNING  ⑦
topology_config {  ⑧
  ...
}
```

① The topology logical plan includes a list of *spouts* elements and a list of *bolts* elements.

② Topology component information, including the component name, a list of key-value configurations, and the component object specification.

③ *spouts* includes a list of *outputs*. *outputs* includes a *stream* and a list of *schema*.

④ Stream information includes the stream ID and parent component name.

⑤ Stream schema is a list of keys/fields defined in the user topology code.

⑥ *bolts* has a list of *inputs* that *spouts* does not have. *inputs* includes *stream* and grouping type.

⑦   The current topology state could be *RUNNING*, *PAUSED*, or *KILLED*.
⑧   The topology logical plan includes a list of key-value configurations.

```
==> PackingPlan:
id: "my-java-topologycaff798c-8c51-483d-9336-815ad55d06b7"
container_plans {  ①
  id: 2  ②
  instance_plans {  ③
    component_name: "exclaim2"  ④
    task_id: 2  ⑤
    component_index: 1  ⑥
    resource {  ⑦
      ...
    }
  }
  ...
  requiredResource {  ⑧
    ...
  }
}
...
```

①   The packing plan is a list of *container_plans*.
②   Container ID, corresponding to Stream Manager ID.
③   *container_plans* includes a list of *instance_plans*.
④   The topology component that this Heron Instance process incarnates.
⑤   The Heron Instance process ID among all Heron Instance processes.
⑥   The topology component "exclaim2" has a list of corresponding Heron Instance processes. *component_index* is the index in this list.
⑦   CPU, RAM, and disk for this Heron Instance process.
⑧   CPU, RAM, and disk for this container.

```
==> PhysicalPlan:
topology {  ①
  ...
}
stmgrs {  ②
  id: "stmgr-1"  ③
  host_name: "vm"  ④
  data_port: 45827  ⑤
  local_endpoint: "/unused"
  cwd: "~/.herondata/topologies/.../my-java-topology"  ⑥
  pid: 24197
  shell_port: 44057  ⑦
  local_data_port: 43367  ⑧
}
...
instances {  ⑨
  instance_id: "container_1_exclaim2_1"  ⑩
  stmgr_id: "stmgr-1"  ⑪
  info {  ⑫
```

```
        . . .
    }
}
. . .
```

①   Similar content to the *topologies* directory. The physical plan includes a
     copy of the logical plan. Both physical and logical plans may change at run-
     time when users issue the `heron update` command. The logical plan here
     is the latest version, while the *topologies* directory keeps its initial version.
②   The physical plan includes a list of *stmgrs*. A Stream Manager represents a
     container since each container has a single Stream Manager.
③   Stream Manager ID, in the form of *stmgr-{CONTAINER_ID}*.
④   Where this Stream Manager process runs.
⑤   Stream Manager listens on this port. The port is used for inter-container
     traffic.
⑥   Container working directory.
⑦   Heron Shell listens on this port.
⑧   Stream Manager listens on this port. The port is used for intra-container
     traffic.
⑨   The physical plan includes a list of *instances*.
⑩   Heron Instance ID, in the form of *container_{CONTAINER_ID}_{COMP
     ONENT_NAME}_{TASK_ID}* .
⑪   What container this Heron Instance runs in.
⑫   Similar identity information in the packing plan *instance_plans* element.

10.3.2   Restart Dead Processes

In Heron, there are two levels of process monitors, as shown in Fig. 10.1. The
first level is Heron Scheduler, which monitors containers; more specifically, Heron
Scheduler monitors Heron Executors. Heron Executor is the first process launched
by the container, and it represents its container. If any Heron Executor dies, Heron
Scheduler restarts or reschedules the dead container.

Heron Executor is the second-level monitor: it starts and monitors the other
processes in the same container. If any monitored process in the same container
dies, Heron Executor restarts the dead process.

A simple experiment to demonstrate Heron Scheduler and Heron Executor
monitoring capability is to kill a Heron Executor or a process in the container
manually by `kill <pid>` and observe the newly started process with a different
PID.

All the process restarting events are logged. The Heron Scheduler log exists at
the *log-files* directory in container 0 with filename format *heron-{TOPOLOGY_
NAME}-scheduler.log.0*. For the topology *my-test-topology-3*, its log is located at
~/.herondata/topologies/local/ubuntu/my-test-topology-3/log-files/.

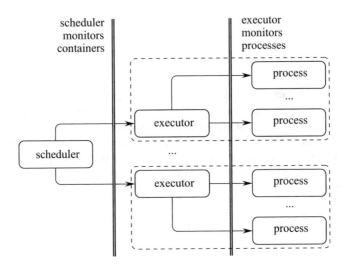

scheduler
monitors
containers

executor
monitors
processes

Fig. 10.1 Two process monitoring levels

Similarly, the Heron Executor log exists at the container working directory with the filename format *heron-executor-{CONTAINER_ID}.stdout*. For the topology *my-test-topology-3*, its log is located at *~/.herondata/topologies/local/ubuntu/my-test-topology-3/*.

10.4 Data Flow

Heron Instance, Stream Manager, and Topology Master collaborate to implement tuple transmission or data flow. Heron Instance executes the spout/bolt code, and the spout/bolt nodes need to communicate with each other. Heron Instances do not connect to each other directly. Instead, a set of Stream Managers connect to Heron Instances offering tuple transmission relay. Stream Managers connect according to the physical plan, which stores all the Stream Managers' metadata. Topology Master, as a well-known authority, maintains the physical plan. The interactions among Heron Instance, Stream Manager, and Topology Master are shown in Fig. 10.2.

The connections among Heron Instance, Stream Manager, and Topology Master are all client–server-type connections. The clients and servers are shown in Fig. 10.3. There are three servers in the figure: *instance server*, *stmgr server*, and *tmaster server*. *instance server* is responsible for all requests from Heron Instances. *stmgr server* is responsible for communication between Stream Managers. *tmaster server* is responsible for all requests from Stream Managers.

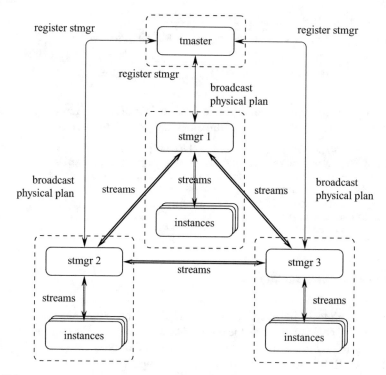

Fig. 10.2 Interaction among tmaster, stmgr, and instance

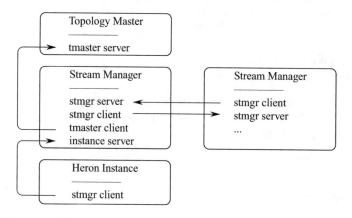

Fig. 10.3 Servers and clients

The client–server connections are all TCP connections with Heron customized protocol. The packet payload format is shown in Table 10.1. To study the Heron data flow, we may capture the communication packets on the server ports and study how they communicate.

Table 10.1 Heron TCP packet payload fields and their sizes

Total length	Protobuf type length	Protobuf type string	REQID	Protobuf data length	Protobuf data
4 bytes	4 bytes	various size	32 bytes	4 bytes	various size

The following sections demonstrate how to interpret the Heron client–server communications. Create the project directory and prepare the scripts:

```
$ mkdir -p ~/heron-proto && cd ~/heron-proto
$ touch capture_packet.py parse_message.py print_table.py
$ tree --charset=ascii ~/heron-proto/
/home/ubuntu/heron-proto/
|-- capture_packet.py
|-- parse_message.py
`-- print_table.py
```

10.4.1 Capture Packets

Heron uses protocol buffers to serialize and deserialize communication data. To parse the TCP payload, we need the protocol buffers compiler. Heron version 0.20.3-incubating-rc7 requires protocol buffers version 3.8.0. protocol buffers project is hosted at GitHub.[1] To install protocol buffers compiler:

```
$ cd ~ && wget https://github.com/protocolbuffers/protobuf/\
> releases/download/v3.8.0/protoc-3.8.0-linux-x86_64.zip && \
> unzip protoc-3.8.0-linux-x86_64.zip -d protoc  ①
$ ~/protoc/bin/protoc --version  ②
```

①   Download protocol buffers compiler.
②   Make sure the `protoc` command is available.

We write three simple scripts to capture the packet, parse the payload, and print on the console. Our scripts depend on several tools to capture packets and format printout. Install the Python tools:

```
$ sudo apt install -y python3-pip
$ sudo pip3 install "netifaces==0.10.9" \  ①
> "scapy==2.4.3" \  ②
> "prettytable==0.7.2"  ③
```

①   Used to list the local network interfaces. We will monitor these interfaces.
②   Scapy is an interactive packet control program. It can manufacture or translate packets of many protocols, catch/send them on the wire, and much more.

---

[1] https://github.com/protocolbuffers/protobuf/releases.

③   PrettyTable is a basic Python library intended to make it quick and easy to visualize tabular data in plain text tables.

Our scripts have three steps, and each step requires some inputs:

1. Capture the TCP packet and extract the payload, which needs to know the port to monitor.
2. Parse the payload, which needs the Heron protocol buffers definition *.proto files. This step calls the **protoc** command to parse the protocol buffers binary.
3. Print the payload content.

Each step is implemented in an individual Python file. Let us first look at *capture_packet.py*. It captures the TCP packets and assembles Heron messages in binary arrays. A TCP packet might not contain a single complete Heron message; thus, *capture_packet.py* has to divide the TCP packet for small Heron messages and assemble large Heron messages from multiple TCP packets.

```python
from netifaces import *
from scapy.all import *
from struct import unpack
from sys import argv

from parse_message import heron_parse

deduplicate = set()   ①
buf_map = dict()   ②

def is_anchor(bin):   ③
  try:
    (total_len,) = unpack('!i', bin[0:4])
    (type_len,) = unpack('!i', bin[4:8])
    type_str = bin[8:8+type_len].decode('ascii')
    if not type_str.startswith('heron.proto.'):   ④
      raise ValueError('unknown proto type prefix %s' % type_str)
    REQID_str = bin[8+type_len:40+type_len].hex()
    (data_len,) = unpack('!i', bin[40+type_len:44+type_len])
    valid = True
  except Exception as e:
    print(e)
    valid = False   ⑤
  return valid

def trial_capture(key, bin, seq):
  ret = '------------\n'+key+'('+str(len(bin))+')\n'
  if (seq in deduplicate):   ⑥
    deduplicate.discard(seq)
    return ret + "duplicated TCP sequence\n"
  else:
    deduplicate.add(seq)

  if is_anchor(bin):
    ret += 'anchor heron packet\n'
    buf_map[key] = bytearray()   ⑦
```

```
if key in buf_map:
  buf_map[key].extend(bin)
  ret += bytes(buf_map[key]).hex()+'\n'
  while len(buf_map[key])>=4+4+32+4:  ⑧
    (total_len,) = unpack('!i', buf_map[key][0:4])
    if len(buf_map[key])<4+total_len:  ⑨
      ret += ('required '+str(4+total_len)+
              '; buffered '+str(len(buf_map[key]))+'\n')
      break
    ret += heron_parse(bytes(buf_map[key][0:4+total_len]))  ⑩
    buf_map[key] = buf_map[key][4+total_len:]  ⑪
else:
  ret += 'unanchored packet\n'
return ret

sniff(iface=list(filter(lambda x: AF_INET in ifaddresses(x),
                 interfaces())),  ⑫
    filter='greater 75 and tcp port '+argv[1],  ⑬
    prn=lambda x: trial_capture(
        x.sprintf('%TCP.sport%->%TCP.dport%'),  ⑭
        bytes(x[TCP].payload), bytes(x[TCP].seq)))
```

①   Scapy detects the packets on the LOOPBACK interface twice—when they "leave" and when they "arrive." We skip every second packet when seeing a TCP sequence number in the denylist `deduplicate`.

②   There are multiple connections on the server port. Each connection is identified by the <TCP.source.port, TCP.destination.port> pair in this experiment. `buf_map` is a map buffering packets for each connection separately.

③   Helper function to tell if a packet is the first packet (anchor packet) of a Heron message by trying to parse the Heron message header.

④   All Heron protocol buffers types start with `heron.proto.`, which can be verified in the source code by `grep '^package' ~/heron/heron/proto/*.proto`.

⑤   If parsing fails in any field of the Heron message header, this packet is assumed not to be an anchor packet.

⑥   When seeing a TCP sequence for the first time, put it in `deduplicate` and continue. When seeing the same TCP sequence in `deduplicate`, flip `deduplicate` to save memory and return.

⑦   When an anchor packet is found, parse all the following TCP packets, thus marking this connection as valid in `buf_map`.

⑧   When the buffered content is long enough, try to parse it in a loop.

⑨   Find the buffered content that is not a complete Heron message and break the loop.

⑩   Invoke `heron_parse()` to parse a binary array to a Heron message. We will see `heron_parse()` soon.

⑪   Remove the parsed content from the buffer.

⑫   Indicate the network interfaces we sniff on. Since the example topology runs in local mode, LOOPBACK may be used, which is included in the list that `interfaces()` returns.

⑬   Filter out the empty payload packets. Sniff on the given port passed in the command-line argument.

⑭   For each captured TCP packet, process it with `trial_capture()` and print the returned string.

NOTE

The maximum transmission unit (MTU) is the maximum sized datagram that can be transmitted through the next network. If a Heron message size is larger than MTU, it will be split into multiple TCP packets.

The MTU for Ethernet, for example, is 1500 bytes. To view the MTU, run these commands:

```
$ ip a | grep mtu
$ ip l | grep mtu
```

The parsing function `heron_parse()` is implemented in *parse_message.py*. It assumes that the input binary array is a valid Heron message.

```
from struct import unpack
from subprocess import PIPE, Popen

from print_table import format_print

def heron_parse(bin):
  (total_len,) = unpack('!i', bin[0:4])  ①
  (type_len,) = unpack('!i', bin[4:8])  ②
  type_str = bin[8:8+type_len].decode('ascii')  ③
  REQID_str = bin[8+type_len:40+type_len].hex()  ④
  (data_len,) = unpack('!i', bin[40+type_len:44+type_len])  ⑤
  data_bin = bin[44+type_len:44+type_len+data_len]  ⑥

  cmd = ('grep -l '+
         type_str.split('.')[-1]+' '  ⑦
         '/home/ubuntu/heron/heron/proto/*.proto\n')  ⑧
  ret = cmd
  p = Popen(cmd, stdin=PIPE, stdout=PIPE, stderr=PIPE,
            shell=True, text=True)
  (PROTO_FILES, err) = p.communicate()  ⑨
  if err != '':
    return ret + err + format_print(
      str(len(bin)), str(total_len), str(type_len),
      type_str, REQID_str, str(data_len), b'')
  ret += PROTO_FILES

  cmd = ('/home/ubuntu/protoc/bin/protoc '+  ⑩
         '--decode='+type_str+' '  ⑪
         '--proto_path=/home/ubuntu/heron/heron/proto/ '+  ⑫
         PROTO_FILES)  ⑬
  ret += cmd
```

```
p = Popen(cmd,
   stdin=PIPE, stdout=PIPE, stderr=PIPE, shell=True)
(data_str, err) = p.communicate(data_bin) ⑭
if err != b'':
  return ret + str(err) + format_print(
    str(len(bin)), str(total_len), str(type_len),
    type_str, REQID_str, str(data_len), data_str)
 return ret + format_print(
   str(len(bin)), str(total_len), str(type_len),
   type_str, REQID_str, str(data_len), data_str)
```

①   The Heron packet header of 4-byte integer specifying the remaining length
      of the packet.
②   The 4-byte integer specifying the string (protocol buffers message type)
      length.
③   A string specifying the protocol buffers message type.
④   REQID of 32 bytes, used as identification for request–response communi-
      cation. For messaging (request without response), the REQID is 0s.
⑤   The 4-byte integer specifying the binary string size.
⑥   Binary string storing a protocol buffers message.
⑦   The pattern to search in the *.proto files.
⑧   Where the *.proto files locate.
⑨   The found .proto file PROTO_FILES containing the target protobufs message
      type.
⑩   Where the protoc command is.
⑪   Target protocol buffers message type.
⑫   What path to search *.proto files.
⑬   What *.proto files to load.
⑭   The decoded protocol buffers string is in the data_str.

Step 3 is separated in the file *print_table.py*. The helper function format_print()
uses the PrettyTable library to display each packet field in Table 10.1. *print_table.py*
looks like:

```
from prettytable import PrettyTable

def format_print(size, header, type_size, type_str,
                 REQID_str, data_size, data_str):
  tbl = PrettyTable(['key', 'value (len in bytes)'])
  tbl.align['key'] = 'r'
  tbl.align['value (len in bytes)'] = 'l'
  tbl.add_row(['payload len', size+'=4+'+header])
  tbl.add_row(['heron len',
               header+'=4+'+type_size+'+32+4+'+data_size])
  tbl.add_row(['proto_type len', type_size])
  tbl.add_row(['proto_type str', type_str])
  tbl.add_row(['REQID', REQID_str])
  tbl.add_row(['proto_data len', data_size])
  tbl.add_row(['proto_data str',
               data_str.decode('ascii').rstrip()])
```

```
    return str(tbl)+'\n'
```

capture_packet.py needs a port number as a launching parameter. The port can be any port that a Heron server listens on. The Heron server port is assigned by Heron Scheduler, which is determined when the topology launches. The server ports are passed to the Heron processes as launching parameters by Heron Executor. To find out the processes launching command, simply use **pstree -al**. The process launching parameters are in key-value pairs. From these key-value pairs, we can find out the target server port easily.

In this section, we run a single topology from Sect. 5.2.2 in the Ubuntu box for the experiment. The following commands show how to find the Stream Manager port and the Topology Master port:

```
$ pstree -al | grep heron-stmgr | grep stmgr-1 | \
> sed 's/^.*--local_data_port/--local_data_port/g' | \
> sed 's/ .*$//g'
--local_data_port=37579  ①

$ pstree -al | grep heron-tmaster | grep topology_id | \
> sed 's/^.*--master_port/--master_port/g' | sed 's/ .*$//g'
--master_port=36959  ②

$ pstree -al | grep MetricsManager | grep metricsmgr-1 | \
> sed 's/^.*--port/--port/g' | sed 's/ .*$//g'
--port=41301  ③
```

①   This example finds the first Stream Manager and displays its server port accepting connections from Heron Instance.

②   This example finds Topology Master and displays its server port responding to Stream Manager connections.

③   This example finds the first Metrics Manager and its server port.

After you collect the ports, run the script with **sudo**, since the network packet capture needs superuser privilege:

```
$ sudo python3 ~/heron-proto/capture_packet.py 37579
```

10.4.2   Communication Primitive

There are several communication primitives in Heron. We picked two typical examples here to demonstrate a common way to interpret them.

Example 1 is a packet from Heron Instance to Stream Manager saying: "I want to send tuples to somebody." Here is the packet:

```
57250->37579(125)
+---------------+-------------------------------------+
|          key  | value (len in bytes)                |
+---------------+-------------------------------------+
|   payload len | 125=4+121                           |
|    heron len  | 121=4+32+32+4+49                    |
| proto_type len| 32                                  |
| proto_type str| heron.proto.system.HeronTupleSet    |
|         REQID | 00000000000000000000000000000000... |
| proto_data len| 49                                  |
| proto_data str| data {                              |
|               |    stream {                         |
|               |       id: "default"                 |
|               |       component_name: "exclaim1"    |
|               |    }                                |
|               |    tuples {                         |
|               |       key: 0                        |
|               |       values: "\003jackson!!\241"   |  ①
|               |    }                                |
|               | }                                   |
|               | src_task_id: 4                      |
+---------------+-------------------------------------+
```

①   Heron uses Kryo[2] for tuple serialization and deserialization.

Example 2 is a packet from Topology Master to Stream Manager saying: "hey,
here is the new physical plan." The new physical plan assignment packet is not often
seen; however, there is a way to trigger it manually. If any container is rescheduled,
the physical plan is changed, resulting in Topology Master sending out a new
physical plan. Assuming we are watching the Topology Master server port, we can
manually restart container 2 in a separate SSH session (see Sect. 8.3.5 for how to
restart a container) to trigger a new physical plan assignment. Since container 0
does not change, the *capture_packet.py* script does not interrupt. Here is the packet:

```
36959->54214(125)
+---------------+------------------------------------------------+
|          key  | value (len in bytes)                           |
+---------------+------------------------------------------------+
|   payload len | 2077=4+2073                                    |
|    heron len  | 2073=4+40+32+4+1993                            |
| proto_type len| 40                                             |
| proto_type str| heron.proto.stmgr.NewPhysicalPlanMessage       |
|         REQID | 00000000000000000000000000000000000000000...  |
| proto_data len| 1993                                           |
| proto_data str| new_pplan {                                    |
|               |    topology {                                  |
|               |       id: "my-java-topologyd0411c5d-74a2-4...  |
|               |       name: "my-java-topology"                 |
|               |       ... ...                                  |
|               |    }                                           |
+---------------+------------------------------------------------+
```

---

[2]https://github.com/EsotericSoftware/kryo.

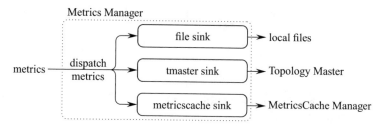

Fig. 10.4 Metrics sinks

10.5   Metrics System

A metric is the measurement of a specific trait of a program's performance or efficiency. Heron metrics include metrics and exceptions, and generalized metrics also include backpressure. We focus on metrics and exceptions in this section.

Each container has a Metrics Manager process responsible for collecting the metrics in the container. The REST processes in the container push the metrics to the Metrics Manager process. The Metrics Manager process owns the metrics sinks configured in *metrics_sinks.yaml*. The metrics sinks are responsible for processing each metrics item, as shown in Fig. 10.4.

There are three often-used metrics sinks: file, Topology Master, and Metrics-Cache Manager. MetricsCache Manager and Topology Master are similar since MetricsCache Manager's goal is to substitute metrics-collector inside Topology Master to simplify the Topology Master functions.

10.5.1   File Sink

With the default configuration, the file sink flushes the metrics to a local file every minute. The file is located in the container's current working directory. The filename is in the format *metrics.json.{metricsmgrid}.{index}*. The file content is in JSON format; for example, to print metrics JSON data, run this command in the current topology working directory:

```
$ python3 -m json.tool --sort-keys < metrics.json.metricsmgr-1.0
[ ①
    . . .
    {
        "context": "default",
        "exceptions": [], ②
        "metrics": { ③
            "metricscache-sink/exceptions-count": "0",
            "metricscache-sink/metrics-count": "48",
            . . .
        },
```

```
        "source": "vm:34763/__metricsmgr__/metricsmgr-1", ④
        "timestamp": 1597008976422
    },
    ...
]
```

①   The file includes a list of metrics messages.
②   A list of exceptions.
③   A map of key-value pairs.
④   Metrics Manager reports metrics to itself.

10.5.2 *MetricsCache Manager Sink and Topology Master Sink*

The MetricsCache Manager sink forwards the metrics to the MetricsCache Manager process in container 0. There is a Metrics Manager process in container 0 as well. MetricsCache Manager, like a normal process, pushes metrics to the local Metrics Manager process as well. MetricsCache Manager is essentially a cache that exposes two services: collecting metrics from Metrics Manager and accepting metrics queries, as shown in Fig. 10.5.

The stats interface of MetricsCache Manager or Topology Master is protobufs over HTTP. MetricsCache Manager provides a tool to query against this stats interface. There are two approaches to run this query tool: from a source code by

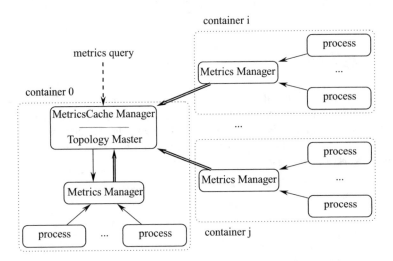

Fig. 10.5 Metrics flow with Topology Master sink or MetricsCache Manager sink

Bazel or from a JAR file by Java. The former approach is similar to what we did in
Sect. 10.2. Here, we demonstrate the latter approach:

```
$ pstree -al | grep MetricsCache | grep stats_port | \
> sed 's/^.*--stats_port/--stats_port/g' | sed 's/ --.*$//g'
--stats_port 44155 ①

$ pstree -al | grep tmaster | grep stats_port | \
> sed 's/^.*--stats_port/--stats_port/g' | sed 's/ --.*$//g'
--stats_port=33739 ②

$ java -cp \
> ~/.heron/lib/metricscachemgr/heron-metricscachemgr.jar \ ③
> org.apache.heron.metricscachemgr.\
> MetricsCacheManagerHttpServer \ ④
> 127.0.1.1:44155 \ ⑤
> exclaim1 \ ⑥
> __jvm-uptime-secs __jvm-memory-used-mb ⑦
endpoint: http://127.0.1.1:44155/stats; component: exclaim1
status {
   status: OK
}
metric {
   instance_id: "container_3_exclaim1_3"
   metric {
      name: "__jvm-uptime-secs"
      interval_values {
         value: "56"
         interval {
            start: 1551545805482
            end: 1551545805482
         }
      }
      ...
   }
   metric {
      name: "__jvm-memory-used-mb"
      interval_values {
         value: "53.0"
         interval {
            start: 1551545805482
            end: 1551545805482
         }
      }
      ...
   }
}
metric {
   instance_id: "container_1_exclaim1_4"
   metric {
      name: "__jvm-uptime-secs"
      ...
   }
```

```
metric {
  name: "__jvm-memory-used-mb"
  ...
}
}
interval: 1551546433
```

(1)   This example finds the query port of MetricsCache Manager.
(2)   This example finds the query port of Topology Master. Both MetricsCache
       Manager and Topology Master ports are ready for metrics queries.
(3)   Where the tools are.
(4)   Main function entry.
(5)   MetricsCache Manager or Topology Master stats host:port. The example
       uses the MetricsCache Manager port.
(6)   Component name in the topology.
(7)   Metrics names, separated by space, for the remaining strings.

MetricsCache Manager or Topology Master accepts a configuration of the list of
metrics in *metrics_sinks.yaml*. Each metric name is prefixed with the aggregation
operation. An excerpt looks like:

```
$ cat ~/.heron/conf/local/metrics_sinks.yaml
...
metricscache-sink: (1)
  ...
  metricscache-metrics-type: (2)
    "__emit-count": SUM (3)
    "__jvm-uptime-secs": LAST
    "__time_spent_back_pressure_by_compid": SUM
    ...
```

(1)   MetricsCache Manager sink configuration object.
(2)   Metrics filters, which means that MetricsCache Manager only accepts these
       metrics and drops the others.
(3)   Metric name and aggregation type. When a query is received, MetricsCache
       Manager aggregates the metrics based on their operation type.

MetricsCache Manager provides the possibility to implement the feedback
system so that the topology can tune or heal itself intelligently. We will cover Health
Manager to close this feedback loop in Sect. 13.1.2.

10.6   Summary

In this chapter, we studied the Heron processes and found the process entry point
for Java, C++, and Python codes. Then, we printed out the Heron state in State
Manager and saw the three plans. We studied the data panel, including Topology

Master, Stream Manager, and Heron Instance, and the communication between them. We captured the communication packet and printed the payload to understand the communication primitives. Finally, we saw how Metrics Manager collects and forwards the metrics to MetricsCache Manager with `metricscache-sink`.

Now, we have a thorough understanding of the Heron system from the topology layer to the Heron layer, as well as from source code to operations. In the next chapter, we will see how to route the metrics flow to a customized sink.

Chapter 11
Extending the Heron Metrics Sink

In the last chapter, we studied the metrics flows as well as two typical metrics sinks. In this chapter, we will try to extend Heron by adding a customized metrics sink. We will first study the Heron metrics series key and out-of-box metrics. Then, we will study what the Heron metrics sink SPI looks like and write an example sink to route the metrics to a MySQL database. The example sink shows the usual way to add a customized metrics sink, which is necessary when Heron topologies run in a data center environment where the central metrics store collects metrics from all running jobs.

11.1 Time Series

Metrics represent how the system performs and usually can be observed through system indicators. Monitoring is gathering, aggregating, and analyzing metrics to improve awareness of system behavior. The monitoring system stores, aggregates, and visualizes metrics from various parts of the system and environment. The monitoring system is also responsible for triggering alerts when the values meet specific conditions.

Metrics are often observed in the form of time series. A time series is a sequence of data points ordered by timestamp. Most usually, a time series is a sampling taken at continuous equal intervals. Thus, it is a sequence of discrete data.

A series key to locate a metric time series is <service, source, metric>, where "service" is the name of a service that the metrics are collected under—think of it as a namespace for metrics—and "source" may be hosts or processes. Mapping to Heron, the series key is <topology name, component and instance IDs, metric name>. An example of a series key is <myTopo, /spout/spout1, __jvm-uptime-secs>, which identifies a time series showing the spout1 JVM running times in seconds.

© The Author(s), under exclusive license to Springer Nature Switzerland AG 2021
H. Wu, M. Fu, *Heron Streaming*, https://doi.org/10.1007/978-3-030-60094-5_11

The metrics observation is essentially a collection of metric time series `Map:`
`<SeriesKey,List<Timestamp|Value>>`. Based on this model, we may construct
our example metrics database as a MySQL database of the following two tables. In
the later example demo, we will route the metrics into this database.

```
CREATE DATABASE IF NOT EXISTS datacentermetricsstore ①
CHARACTER SET = 'ascii' COLLATE = 'ascii_general_ci';

CREATE TABLE IF NOT EXISTS `serieskey` ( ②
  `seriesid` SMALLINT NOT NULL AUTO_INCREMENT PRIMARY KEY, ③
  `toponame` TINYTEXT NOT NULL,
  `componentinstance` TINYTEXT NOT NULL,
  `metircname` TINYTEXT NOT NULL,
  UNIQUE KEY `seriesidx` (`toponame`(200),
                          `componentinstance`(200),
                          `metircname`(200)) ④
);

CREATE TABLE IF NOT EXISTS `metricval` ( ⑤
  `seriesid` SMALLINT NOT NULL,
  `ts` TIMESTAMP NOT NULL DEFAULT CURRENT_TIMESTAMP,
  `val` DOUBLE NOT NULL,
  FOREIGN KEY (seriesid) REFERENCES serieskey(seriesid) ⑥
);
```

①   The database to collect metrics from all jobs in a data center.
②   The table to store series keys. Each key corresponds to a metric time series
        in the table `metricval`.
③   The series key ID to identify the series key. This field is uniquely autogen-
        erated as the primary key of the table.
④   Each series key has three tags. The three tags uniquely identify a time series
        as well as its series key. Since the index requires a fixed field size, the
        first 200 of 256 bytes—TINYTEXT is a string that can store up to 255
        characters—are chosen to form the unique index `seriesidx`.
⑤   The table to store time series data points.
⑥   The column `seriesid` reference to table `serieskey`'s primary key, which
        guarantees data validation.

NOTE

To install MySQL on a fresh machine, use the following commands:

```
$ sudo apt install -y mysql-server
$ sudo mysql ①
mysql> CREATE USER 'username'@'localhost'
    -> IDENTIFIED BY 'password';
mysql> GRANT ALL PRIVILEGES ON *.*
    -> TO 'username'@'localhost' WITH GRANT OPTION;
mysql> exit
$ mysql -u username -p
```

① MySQL, rather than with a password, authenticates the root user using the auth_socket plugin by default.

11.2 Metrics Category

There are three essential categories of metrics in Heron:

Work metrics
Include throughput or counter, acked and failed tuple count, latency or processing time, etc.
Resource metrics
Include CPU and memory utilization, backpressure time showing saturation, etc. All metrics reach Metrics Manager and then are filtered by metrics sinks.
User-defined metrics
Can be added in the topology code with the Heron API in a similar way as the Storm API does. Moreover, Heron provides a new metric type called GlobalMetrics, which is essentially a MultiCountMetric singleton exposing a simple, globally available counter for Heron jobs. Anywhere in the execution of the Heron job, topology writers can put GlobalMetrics.incr("mycounter") to add a counter. The counters will be named __auto__/mycounter (note the __auto__ prefix). There is no need to declare the counter before explicitly. If the counter does not exist, it will be created. The creation is lazy, which means that unless the counter is not available, it is counted as 0. The ExclamationTopology example topology shows the GlobalMetrics counter in ExclamationBolt.execute(). We will observe __auto__/selected_items metric in Sect. 11.4.3.

11.3 Customize a Metrics Sink

Each Heron container has a single Metrics Manager process responsible for collecting metrics from all processes in the container. The Metrics Manager process is constructed following the plug-in pattern, and you can implement your metrics sink and plug it to the Metrics Manager process.

Heron comes equipped out of the box with three metrics sinks that you can apply for a specific topology. The code for these sinks may be helpful for implementing your own sinks [1]:

- PrometheusSink sends metrics to a specified path in the Prometheus instance.
- GraphiteSink sends metrics to a Graphite instance according to a Graphite prefix.

- ScribeSink sends metrics to a Scribe instance according to a Scribe category and namespace.

11.3.1 Metrics Sink SPI

The metrics sink SPI is in Java. The major interface you must implement is IMetricsSink, which has the following methods:

init()
> Two parameters are passed to this method. The first parameter is a configuration map constructed from the *metrics_sinks.yaml* file. This YAML file is organized by metrics sinks. Each metrics sink has a map and is isolated from the other metrics sinks. The second parameter is SinkContext, which is essentially another key-value map, including the topology name, Metrics Manager ID, metrics sink ID, etc. We distinguish the two key-value map parameters carefully. The first configuration map is populated from the YAML file and would not be changed anymore, and the second parameter, SinkContext, is populated at runtime. Many init() of SPI interfaces accept two map parameters like this, for example, the ILauncher and IScheduler in Chap. 12.

processRecord()
> This method accepts a single parameter in the MetricsRecord type. A MetricsRecord object is an immutable snapshot of metrics and exception log with a timestamp and other metadata. Heron Instance captures exceptions thrown from the spout and bolt logic and reports them to Metrics Manager. The exceptions in the user's spout and bolt logic are regarded as user-defined events; thus, they are handled together like metrics.

flush()
> This method is called at an interval according to the configuration. The item flush-frequency-ms of each metrics sink in the *metrics_sinks.yaml* file indicates this interval. If flush-frequency-ms is not set, flush() is never called.

close()
> This method frees any allocated resources.

11.3.2 Metrics Sink Configuration

The *metrics_sinks.yaml* file indicates what metrics sinks should be launched and how they are configured. At the top of that file is the sinks list that includes the sinks you want to use. Here is an example:

```
sinks:
  - file-sink
  - metricscache-sink
  - prometheus-sink
```

For each sink, you are required to specify the following [1]:

```
class
```
> The Java class name of your implementation of the `IMetricsSink` interface, for
> example, `org.apache.heron.metricsmgr.sink.tmaster.TMasterSink`.

```
flush-frequency-ms
```
> The frequency (in milliseconds) at which the `flush()` method is called in your
> implementation of `IMetricsSink`.

```
sink-restart-attempts
```
> Indicates how many times a sink will be restarted if it encounters exceptions and
> dies. A Metrics Manager process usually is configured with multiple metrics sink
> objects. Some metrics sinks may go wrong at runtime due to some reason. This
> configuration is introduced to mitigate the impact that broken metrics sinks have
> on other good metrics sinks. The default is 0, which means the sink is gone once
> any exception is thrown. If you set it to −1, the sink will try to restart forever.

Some sinks require other configurations that should be specified here together. All
configurations will be passed to the sink `init()` method as an unmodifiable map.

11.4 MySQL Metrics Sink

The previous sections discussed the concepts and steps to write a customized metrics
sink. This section will show an example metrics sink to write selected metrics to a
MySQL database. The demonstration project starts with the clean source code:

```
$ cd ~ && tar -xzf 0.20.3-incubating-rc7.tar.gz && \
> mv incubator-heron-0.20.3-incubating-rc7/ \
> heron-metrics-sink/ && cd heron-metrics-sink/
```

To connect to a MySQL database, a MySQL connector or MySQL JDBC
connector is necessary. This dependency should be added to the *WORKSPACE* file
and be referenced in the Metrics Manager *BUILD* file as Bazel requires.

```
maven_install(
    name = "maven",
    artifacts = [
        "mysql:mysql-connector-java:8.0.21",  ①
        ...
```

①   Connector/J version 8.0 is a Type 4 pure Java JDBC 4.2 driver. It is
    compatible with all the functionality of MySQL 5.5, 5.6, 5.7, and 8.0.

Update *maven_install.json* to refresh the newly added dependency:

```
$ bazel run @unpinned_maven//:pin
```

Reference this dependency in `heron/metricsmgr/src/java/BUILD:deps` by adding:

```
deps = [
  ...
  "@maven//:mysql_mysql_connector_java",
]
```

11.4.1   Implement IMetricsSink

The key methods of the `IMetricsSink` interface are implemented in a new file *heron/metricsmgr/src/java/org/apache/heron/metricsmgr/sink/MysqlSink.java*, with the following skeleton:

```
public class MysqlSink implements IMetricsSink { ①
  private SinkContext ctx; ②
  private Connection conn = null; ③
  private MetricsFilter mf; ④

  @Override
  public void init(Map<String, Object> m, SinkContext c) {...}⑤

  @Override
  public void processRecord(MetricsRecord record) {...} ⑥

  @Override
  public void flush() {...} ⑦

  @Override
  public void close() {...} ⑧
}
```

①   The customized sink implements `IMetricsSink`.

②   `SinkContext` indicates topology metadata, such as topology name.

③   The connection to the MySQL database, with which all database operations are performed.

④   The filter representing the allowlist configuration in the YAML file *metrics_sinks.yaml*.

⑤   Do initialization tasks in the function, including MySQL connection initialization and loading the configuration from the YAML file.

⑥   Process a MetricsRecord object in this function. We will send the metrics to the MySQL database in this function.

⑦   Commit the database transactions in this function.

⑧   Free resources in this function.

Let us first look at the initialization function `init()`:

```java
public void init(Map<String, Object> m, SinkContext c) {
  this.ctx = c;
  try {
    this.conn = DriverManager.getConnection(
      (String) m.get("mysql-site"), (String) m.get("mysql-user"),
      (String) m.get("mysql-pass")); ①
    conn.setAutoCommit(false); ②
  } catch (SQLException e) {
    e.printStackTrace();
  }
  this.mf = new MetricsFilter();
  ((List<String>) m.get("mysql-metric-name-prefix")).stream()
    .forEach(x -> mf.setPrefixToType(x,
      MetricsFilter.MetricAggregationType.UNKNOWN)); ③
}
```

①   Read three configuration items—`mysql-site`, `mysql-user`, and `mysql-pass`—in the YAML file and construct the connection to the database with MySQL connector's help.

②   Since `flush()` commits batch operations, the auto-commit is turned off.

③   Read the configuration `mysql-metric-name-prefix` in the YAML file, which returns a list, and construct the `MetricsFilter` object that is used to filter metrics in `processRecord()`. `MetricsFilter` is essentially a `Map` whose key is the metric name prefix and whose value is the corresponding aggregation type. The aggregation is not enforced in our example; thus, the `UNKNOWN` type is used.

`processRecord()` does the actual work to insert metrics into the database:

```java
public void processRecord(MetricsRecord record) {
  String toponame = ctx.getTopologyName();
  String source = record.getSource();
  for (MetricsInfo mi: mf.filter(record.getMetrics())) { ①
    String metricname = mi.getName();
    String value = mi.getValue(); ②
    try (Statement stmt = conn.createStatement()) { ③
      String sql1 = String.format(
        "INSERT IGNORE INTO serieskey " +
        "(toponame, componentinstance, metircname) " +
        "VALUES ('%s', '%s', '%s');",
        toponame, source, metricname); ④
      stmt.executeUpdate(sql1);
      String sql2 = String.format("INSERT INTO metricval " +
        "(seriesid, ts, val) " +
        "SELECT seriesid, CURRENT_TIMESTAMP(), %s " +
        "FROM serieskey WHERE toponame='%s' " +
        "AND componentinstance='%s' AND metircname='%s';",
        value, toponame, source, metricname); ⑤
      stmt.executeUpdate(sql2);
    } catch (SQLException e) {
```

```
      e.printStackTrace ();
    }
  }
}
```

① `MetricsRecord` contains a list of metrics and a list of exceptions. Our
    example focuses on the metrics; the exceptions can be handled similarly and
    are not discussed here. While iterating the metrics list, the filter is applied.
② Until now, the series key of < `toponame`, `source`, `metricname` > and data
    point value are collected, ready to publish the metric to a MySQL database.
③ The try-with-resources statement ensures that the resource `stmt` is closed at
    the end of the statement.
④ This is the first step to publish the metric to the database. It assures that the
    series key exits in the table `serieskey`. An example of SQL looks like:

```
INSERT IGNORE INTO serieskey
(toponame, componentinstance, metircname)
VALUES ('myTopoEt',
        'vm:5/word/container_1_word_5',
        '__jvm-process-cpu-load');
```

If the series key already exists, no change will be done.

⑤ This is the second step to publish the metric to the database. It first finds the
    series key in the table *serieskey*, its existence is guaranteed in the first step
    SQL, and then it inserts the data point of < `CURRENT_TIMESTAMP()`, value
    > into the table `metricval`. The timestamp member in the `MetricsRecord`
    is not adopted because (a) the default metrics collection interval is 1 min,
    which is not time-critical and (b) the timestamp member is set as the metrics
    packet arriving timestamp at Metrics Manager; in other words, it is not the
    accurate metric birth time. An example of SQL looks like:

```
INSERT INTO metricval (seriesid, ts, val)
SELECT seriesid, CURRENT_TIMESTAMP(), 0.6979002704728091
FROM serie skey
WHERE toponame='myTopoEt'
AND componentinstance='vm:5/word/container_1_word_5'
AND metircname='__jvm-process-cpu-load';
```

`flush()` tries to commit the transactions at the configured interval. If an exception
is thrown, try to roll back. The flush interval is indicated by `flush-frequency-ms`
in the YAML file:

```
public void flush() {
  try {
    conn.commit();
  } catch (SQLException e) {
    try {
      conn.rollback();
    } catch (SQLException e1) {
      e1.printStackTrace();
    }
```

```
    }
}
```

The last method is `close()` that closes the MySQL connection:

```
public void close() {
  try {
    conn.close();
  } catch (SQLException e) {
    e.printStackTrace();
  }
}
```

11.4.2 Configure the Metrics Sink

The configuration is divided into two levels. First, let Metrics Manager know the new `mysql-sink`; second, configure the `mysql-sink` metrics sink. The following example shows how to configure the metrics sink in the *metrics_sinks.yaml* file:

```
sinks:  ①
  - mysql-sink  ②

mysql-sink:  ③
  class: "org.apache.heron.metricsmgr.sink.MysqlSink"  ④
  flush-frequency-ms: 120000  ⑤
  sink-restart-attempts: 3  ⑥
  mysql-site: "jdbc:mysql://127.0.0.1:3306\
/datacentermetricsstore?serverTimezone=America/Los_Angeles"  ⑦
  mysql-user: "username"
  mysql-pass: "password"  ⑧
  mysql-metric-name-prefix:  ⑨
    - __emit-count  ⑩
    - __jvm-process-cpu-load  ⑪
    - __auto__  ⑫
```

①   By adding the new metrics sink `mysql-sink` in the list under the item `sinks`, Metrics Manager is aware of the new metrics sink.

②   Add the new metrics sink to the list.

③   The item for the new metrics sink configurations.

④   Tells which class implements the `IMetricsSink` interface.

⑤   Tells how often `flush()` is called.

⑥   Tells how many times to restart this metrics sink if it fails. If the value is negative, Metrics Manager tries to restart it forever when seeing its failure.

⑦   The MySQL connection string. It indicates the database host, port, and database name. The time zone is necessary to mitigate the connection exception.

⑧   MySQL connection authentication pair: username and password. Change to your username and password.

	(9)	This item is used to filter metrics.
	(10)	Work metric indicating how many tuples are emitted in a metrics interval time window. The default is a 1-min window.
	(11)	Resource metric indicating the recent JVM CPU load.
	(12)	User-defined event metric, which starts with the prefix __auto__.

11.4.3 Observe Metrics

Before running the topology, make sure you can access the database with username and password in the YAML file and make sure that the tables exist, as the SQL commands shown in Sect. 11.1.

Follow the steps in Chap. 3 to compile the source code and submit an example topology. Then, check the database tables to view the collected metrics:

```
$ mysql -u username -p datacentermetricsstore -e \
> "SELECT * FROM serieskey"
Enter password:
+-----------+-----------+----------------------+---------------+
| seriesid  | toponame  | componentinstance    | metircname    |
+-----------+-----------+----------------------+---------------+
|         1 | myTopoEt  | vm:5/word/contain... | __jvm-proc... |
|         2 | myTopoEt  | vm:44981/__metric... | __jvm-proc... |
...

$ mysql -u username -p datacentermetricsstore -e \
> "SELECT * FROM metricval WHERE seriesid=10 \
> ORDER BY ts DESC LIMIT 10"
Enter password:
+-----------+----------------------+-----+
| seriesid  | ts                   | val |
+-----------+----------------------+-----+
|        10 | 2020-08-10 01:36:28  | 41  |
|        10 | 2020-08-10 01:36:28  | 41  |
|        10 | 2020-08-10 01:34:28  | 29  |
...
```

The above excerpt shows that the metric source is constructed in the format host:port/component_name/instnance_id. For Heron Instance, its port is not available; thus, its task ID is used.

11.5 Summary

In this chapter, we described metrics as time series, based on which we constructed the database schema for our MySQL metrics sink. We also discussed three metrics types and picked three representative metrics for a MySQL metrics sink. We studied the metrics sink SPI and how to configure a metrics sink in a YAML file.

We studied step by step how to write a MySQL metrics sink to publish the metrics to a MySQL database. The MySQL metrics sink example demonstrated how to implement the `IMetricsSink` interface and what the configuration in the YAML file looks like. After studying and writing the MySQL metrics sink, we can follow the same steps to write any new metrics sinks.

Reference

1. Implementing a custom metrics sink. https://heron.incubator.apache.org/docs/extending-heron-metric-sink. Visited on 2020-07

Chapter 12
Extending Heron Scheduler

To run a Heron topology, you will need to set up a scheduler that is responsible for topology management. One scheduler manages only one topology. Heron currently supports the following schedulers out of the box (installed in the directory *~/.heron/lib/scheduler/*):

- Kubernetes
- Apache Hadoop YARN
- Apache Aurora
- Local scheduler
- HashiCorp Nomad
- Apache Marathon
- Apache Mesos
- Slurm
- Null scheduler

If you would like to run Heron on a not-yet-supported resource pool, such as Amazon ECS, you can create your scheduler using Heron's SPI, as detailed in Sect. 12.1. Java is currently the only supported language for custom schedulers. This may change in the future. In this chapter, we will go through an example of how to write a *timeout* scheduler in Sect. 12.2.

12.1 Scheduler SPI

Creating a custom scheduler involves implementing two pivotal Java interfaces:

IScheduler
  Defines how to manage containers and the topology life cycle.
ILauncher
  Defines how to launch the scheduler.

H. Wu, M. Fu, *Heron Streaming*, https://doi.org/10.1007/978-3-030-60094-5_12

Besides both IScheduler and ILauncher, there are several related interfaces, including:

IScalable
Defines how to scale a topology up and down by adding and removing containers.
IPacking
Defines the algorithm used to generate a packing plan for a topology.
IRepacking
Defines the algorithm used to change the packing plan during the scaling process.
IUploader
Uploads the topology to a shared location accessible to the runtime environment of the topology.

During the topology submission process, the involved interfaces include ILauncher, IPacking, IUploader, and IScheduler. During topology runtime, to respond to a topology life cycle state change, the involved interfaces include IScalable, IRepacking, and IScheduler. The IScheduler interface is involved in both situations.

12.1.1 ILauncher and IScheduler Work Together

When a user starts a Heron CLI process to submit a topology, the Heron CLI first calculates a packing plan by invoking IPacking, then uploads the topology file to a shared store by invoking IUploader and calls ILauncher to start the scheduler process, which finally calls IScheduler to run containers.

There are two scheduler modes—service and library. In service mode, the default mode, the scheduler runs as a service process or daemon. There must be a place to run this scheduler service daemon, which is usually not the user's machine where the Heron CLI process runs. The scheduler service usually sits with Topology Master in container 0, but it can be run anywhere as long as the following criteria are satisfied:

- The scheduler process lives the entire topology life cycle since it has to manage the topology life cycle.
- The scheduler process is accessible by the Heron CLI because the Heron CLI has to launch this scheduler process.
- The scheduler process can manage containers to fulfill start, kill, and restart container functions.

If service mode is not applicable in some scenarios, the library mode is an alternative. In library mode, there is no scheduler process or daemon; instead, IScheduler is called in the Heron CLI process directly as a library. Due to the lack of a live process to maintain the state between the IScheduler functions, the Heron CLI process has to find a way to transfer the knowledge between the IScheduler functions, for example, by implicit agreement among the IScheduler functions or by delegating to a third agent.

12.1.2   ILauncher

Let us think of a typical scenario as an example—default service mode with the scheduler process inside container 0. Since the scheduler process runs in container 0, the Heron CLI process has to manage container 0 at least. To make the Heron resource pool agnostic, ILauncher is introduced to spawn container 0 and run the scheduler process in it. The ILauncher consists of the following important methods:

initialize()
>   This method accepts two parameters, config and runtime, of the same type org.apache.heron.spi.common.Config. The config parameter contains the information before the Heron CLI runs, including the configuration in YAML files and command-line parameters. runtime contains the information related to topology, uploader, packing plan, etc., generated during the Heron CLI runs.

launch()
>   This method accepts a packing plan and starts a scheduler feeding it the packing plan. Once this function returns successfully, the Heron CLI will terminate and the launch process succeeds.

12.1.3   IScheduler

To make Heron run on multiple resource pools and hide the resource pool details, IScheduler is introduced. It consists of the following important methods:

onSchedule()
>   This method will be called after initialize(). It is responsible for grabbing resources to launch Heron Executor and make sure they get launched.

onKill(), onRestart(), and onUpdate()
>   These methods are called by the scheduler service in the service mode or by the Heron CLI in the library mode. They are responsible for changing the topology life cycle state by manipulating Heron Executor in the resource pool.

12.2   Timeout Scheduler

The previous section discussed the basic concepts and steps to write a customized scheduler. Now it is time to write some code. This section shows an example *timeout* scheduler. The *timeout* scheduler is similar to the *local* scheduler but with a timeout threshold. Heron assumes the topology is a long-running job and stops only when the **heron kill** command is issued. The example *timeout* scheduler changes this assumption a bit. It stops when **heron kill** is issued or a timeout occurs, whichever comes first. To make the example comprehensive and straightforward to demonstrate the scheduler customization, the example *timeout* scheduler implements only two methods, onSchedule() and onKill(), in both service and library modes.

The launcher is responsible for launching the scheduler. It can launch a scheduler in either service mode or library mode. A switch is necessary to indicate to the launcher which scheduler mode should be used. A Boolean flag, which may be fed by the user in the command line or in the YAML file as a configuration parameter, can fulfill the role of a switch easily.

To organize two scheduler modes, two scheduler classes—each implementing a single mode—is a simple and clear design. The two scheduler classes may share some common operations which can be put in an abstract class as their parent. Thus, let the parent class TimeoutSchedulerAbstract implement initialize(), close(), getJobLinks(), onRestart(), and onUpdate(), while the subclasses TimeoutSchedulerService and TimeoutSchedulerLibrary implement only onSchedule() and onKill().

Follow the similar commands in Sect. 11.4 to create a project directory *~/heron-scheduler/*. Since the *timeout* scheduler is similar to the *local* scheduler, the *local* scheduler code should be studied and may be reused. A new code directory named *timeout* can be created besides the *local* scheduler directory as *~/heron-scheduler/heron/schedulers/src/java/org/apache/heron/scheduler/timeout*. Create four Java files in the directory: *TimeoutLauncher.java*, *TimeoutSchedulerAbstract.java*, *TimeoutSchedulerLibrary.java*, and *TimeoutSchedulerService.java*. Then find the Bazel target local-scheduler-unshaded in the *heron/schedulers/src/java/BUILD* file and append the four Java files to this target:

```
java_binary(
    name='local-scheduler-unshaded',  ①
    srcs = glob(["**/local/*.java"]) +
           glob(["**/timeout/*.java"]),  ②
    deps = scheduler_deps_files,
)
```

①   This Bazel target builds the *local* scheduler binary.
②   Add the *timeout* scheduler code to the *local* scheduler binary so that they can be invoked as the *local* scheduler is invoked.

The *timeout* launcher and scheduler classes should be configured so that the Heron CLI can invoke them. Since the *local* scheduler binary is reused, its YAML configuration *heron/config/src/yaml/conf/local/scheduler.yaml* can be updated to the new launcher and scheduler:

```
heron.scheduler.timeout.duration: "60"  ①
heron.class.launcher: >
  org.apache.heron.scheduler.timeout.TimeoutLauncher  ②
heron.scheduler.is.service: True  ③
heron.class.scheduler: >
  org.apache.heron.scheduler.timeout.TimeoutSchedulerService  ④
```

```
#heron.scheduler.is.service: False
#heron.class.scheduler: >
#  org.apache.heron.scheduler.timeout.TimeoutSchedulerLibrary⑤
```

- ①   The timeout threshold is 60 s.
- ②   Replace the local launcher with the timeout launcher.
- ③   The Boolean flag to indicate to the launcher which scheduler mode is used.
- ④   Replace the *local* scheduler with the *timeout* scheduler.
- ⑤   The scheduler class for the library mode.

The Heron CLI supports `--config-property key=value` to override the configuration pairs in the YAML files. Thus the above configuration may be fed in the command line, while the YAML file remains unchanged.

12.2.1 Timeout Launcher

The launcher object is born when its `initialize()` is called and dies when `close()` is called. When the launcher object is alive, its `launch()` method is called once when the scheduler should be launched. The launcher object lives in the Heron CLI process and dies when the Heron CLI process finishes the **heron submit** command. The launcher implementation is in *TimeoutLauncher.java* that has the following skeleton:

```
public class TimeoutLauncher implements ILauncher {  ①
  @Override
  public void initialize(Config config, Config runtime) {...}②

  @Override
  public void close() {}  ③

  @Override
  public boolean launch(PackingPlan packing) {...}  ④

  private void prepareDirectroy() {...}  ⑤
  private boolean startSchedulerAsyncProcess() {...}  ⑥
  private boolean onScheduleAsLibrary(PackingPlan packing)  ⑦
}
```

- ①   The launcher implements the `ILauncher` interface.
- ②   `initialize()` keeps local copies of given configuration maps.
- ③   `close()` is empty.
- ④   `launch()` first prepares the local directory as containers; then, it launches the scheduler in either service mode or library mode according to the configuration in the YAML file.
- ⑤   Prepare a directory serving as container root directories.
- ⑥   Scheduler service mode implementation.
- ⑦   Scheduler library mode implementation.

The launcher object has the opportunity of keeping the configuration in the `initialize()` method for later use in `launch()`. One particular configuration for both the *local* scheduler and the *timeout* scheduler is the working directory because they both isolate the topologies by assigning them different working directories. Since no resource is allocated in `initialize()`, `close()` is empty. `initialize()` looks like:

```
public void initialize(Config config, Config runtime) {
  this.conf = Config.toLocalMode(config);
  this.runt = Config.toLocalMode(runtime);
  this.workingDirectory = LocalContext.workingDirectory(conf);
}
```

`launch()` does the actual work to launch the scheduler:

```
public boolean launch(PackingPlan packing) {
  prepareDirectroy();        ①
  if (conf.getBooleanValue(Key.SCHEDULER_IS_SERVICE)) {   ②
    return startSchedulerAsyncProcess();
  } else {
    return onScheduleAsLibrary(packing);    ③
  }
}
```

①   Although the topology working directory is available and set in `initialize()`, some work has to be done to make the directory ready as a container. See Sect. 12.2.1.1 for details.

②   Read the configuration and branch to either service mode or library mode. The configuration is in the YAML file *heron/confg/src/yaml/conf/local/scheduler.yaml*.

③   The service mode does not need the packing plan because the new scheduler process will get the packing plan from State Manager. In contrast, the library mode needs the packing plan, which should be passed to the scheduler's `onSchedule()`.

12.2.1.1   Prepare Container

The *timeout* scheduler follows the same way as the *local* scheduler does to organize the directory or container. The scheduler process runs in container 0 with Topology Master. Moreover, the other containers share the same directory with container 0. Thus the working directory serves as both container 0 and the regular containers.

If the scheduler process is not put in container 0, the container preparation may be postponed to the scheduler's `onSchedule()`. Similarly, for library mode without the scheduler process, the container preparation should be in the scheduler's `onSchedule()`. For the *timeout* scheduler to cover both service and library modes, the container preparation is put inside the launcher's `launch()`.

A container is ready as long as it contains two packages—the Heron core and the user's topology JAR file. The Heron core package includes the code to run Heron processes, while the topology JAR file includes the code to run user logic. Only if both packages exist can the topology run successfully. `prepareDirectory()` makes the directory ready as a container to launch the scheduler process:

```java
private void prepareDirectory() {
  SchedulerUtils.createOrCleanDirectory(workingDirectory);   ①

  Path heronCore =
    Paths.get(LocalContext.corePackageDirectory(conf));   ②
  Path heronCoreLink =
    Paths.get(workingDirectory, "heron-core");   ③
  try {
    Files.createSymbolicLink(heronCoreLink, heronCore);   ④
  } catch (IOException e) {
    e.printStackTrace();
  }

  String topoURI =
    Runtime.topologyPackageUri(runt).toString();   ⑤
  String topoDest =
    Paths.get(workingDirectory, "topology.tar.gz")
        .toString();   ⑥
  SchedulerUtils.extractPackage(
    workingDirectory, topoURI, topoDest, true, true);   ⑦
}
```

①   Make sure that the directory exists and clean the directory.

②   The Heron core directory of the Heron CLI installation, usually *~/.heron/dist/heron-core*.

③   The container has to own a copy of the Heron core. The Heron core includes the binaries, such as *tmaster* and *stmgr*, and libraries, such as *statemgr*, to run the topology components. The Heron core directory inside the container is conventionally named *heron-core*.

④   For both *local* and *timeout* schedulers, a symbolic link points to the *heron-core* directory of the Heron CLI installation to avoid duplicate *heron-core* copies.

⑤   For both *local* and *timeout* schedulers, the `LocalFileSystemUploader` uploader puts the topology JAR file in the shared store, usually the directory *~/.herondata/repository/topologies/local/[user_name]/[topology_name]*.

⑥   The topology JAR file is named *topology.tar.gz* inside the container.

⑦   Download the JAR with the `curl` command from the shared directory to the container directory and extract the package content with the `tar` command.

12.2.1.2 Launch by Service

To start the scheduler service process, the launcher usually first starts container 0 and then starts the scheduler process. The YARN scheduler works in this way. For both *local* and *timeout* schedulers, the first step—starting container 0—is omitted. `startSchedulerAsyncProcess()` calculates the scheduler command and starts it in a separate process:

```
private boolean startSchedulerAsyncProcess() {
  List<Integer> port = IntStream
    .generate(() -> SysUtils.getFreePort())
    .limit(SchedulerUtils.PORTS_REQUIRED_FOR_SCHEDULER).boxed()
    .collect(Collectors.toList()); ①
  String[] schedulerCmd =
    SchedulerUtils.schedulerCommand(conf, runt, port); ②
  ShellUtils.runASyncProcess(schedulerCmd,
    new File(workingDirectory), null); ③
  return true; ④
}
```

①　Occupy the ports for scheduler process use. This is for the *local* and *timeout* schedulers only. For the other container pools, such as Aurora, the ports are assigned to the container by Aurora.

②　Calculate the scheduler process command. It usually looks like:

```
java -cp \
~/.heron/lib/scheduler/*:\
~/.heron/lib/packing/*:\
~/.heron/lib/statemgr/* \
org.apache.heron.scheduler.SchedulerMain \
--cluster local --role ubuntu --environment default \
--topology_name myTopoEt \
--topology_bin heron-api-examples.jar \
--http_port <port>.
```

③　Run the calculated command in a separate process. The new process will still be running after the Heron CLI process quits.

④　`launch()` returns true, meaning a successful launch.

12.2.1.3 Launch by Library

For the library mode, a scheduler object has to be initialized, and then its `onSchedule()` is called to launch the topology. The normal method invocation order on the scheduler object is `initialize()`, `onSchedule()`, and `close()`:

```
private boolean onScheduleAsLibrary(PackingPlan packing) {
  try {
    IScheduler scheduler = (IScheduler) ClassLoader
      .getSystemClassLoader()
      .loadClass(conf.getStringValue(Key.SCHEDULER_CLASS))
      .newInstance(); ①
```

```
    scheduler.initialize(conf, LauncherUtils.getInstance()
        .createConfigWithPackingDetails(runt, packing)); ②
    return scheduler.onSchedule(packing); ③
  } catch (ClassNotFoundException |
          IllegalAccessException |
          InstantiationException e) {
    e.printStackTrace();
  }
  return false; ④
}
```

① Read the configuration from the YAML file *heron/confg/src/yaml/conf/local/scheduler.yaml*, find the scheduler class, and then generate the object by Java reflection mechanism.
② Call the first method in the scheduler invocation order initialize().
③ Call the second method in the scheduler invocation order onSchedule(). close() is omitted since it is empty.
④ If any exception is captured, return false, indicating the launch failed.

12.2.2 Timeout Scheduler

The scheduler is responsible for starting and monitoring the Heron Executor process in the container. The *timeout* scheduler example focuses on the onSchedule() and onKill() methods, and lets onRestart(), onUpdate(), and getJobLinks() be empty. Its initialize() and close() work the same way as TimeoutLauncher does. The empty and the shared methods are put in the abstract parent class in *TimeoutSchedulerAbstract.java*.

12.2.2.1 Abstract Parent Class

One shared operation for both service and library modes is to generate the Heron Executor command, as shown in getExecutorCommand():

```
protected String[] getExecutorCommand(int shardId) {
  Map<ExecutorPort, String> port =
    ExecutorPort.getRequiredPorts().stream().collect(
      Collectors.toMap(ep -> ep,
        ep -> Integer.toString(SysUtils.getFreePort())));  ①

  List<String> executorCmd = new ArrayList<>();
  executorCmd.add("timeout"); ②
  executorCmd.add(conf.getStringValue(
    "heron.scheduler.timeout.duration", "120"));  ③
  executorCmd.add(Context.executorBinary(conf)); ④
  executorCmd.addAll(
    Arrays.asList(SchedulerUtils.executorCommandArgs(
```

```
         conf, runt, port, Integer.toString(shardId))))); (5)
   return executorCmd.toArray(new String[0]);
}
```

(1)   Occupy the ports for the Heron Executor process.

(2)   `timeout` *[OPTION] DURATION COMMAND [ARG]...* starts the *COMMAND* and kills it if still running after *DURATION* seconds. The *timeout* scheduler uses a timeout to expire the Heron Executor process.

(3)   *DURATION*: The timeout duration is read from the configuration. If the configuration is not found, use the default value of 2 min.

(4)   *COMMAND*: The Heron Executor binary, which looks like *./heron-core/ bin/heron-executor*.

(5)   *ARG*: The parameters to the Heron Executor binary. The `shardId` differentiates the Heron Executor processes. The whole command looks like `timeout 120 ./heron-core/bin/heron-executor --shard=...`.

12.2.2.2 Service Mode

The service mode scheduler, implemented in *TimeoutSchedulerService.java*, tracks the Heron Executor processes that it starts. Thus a hash set `processes` member is included in the scheduler. Meanwhile, the scheduler watches the Heron Executor process state using an `ExecutorService`. The typical service mode scheduler monitors and restarts Heron Executor processes if any die. The simple *timeout* scheduler monitors Heron Executor processes and quits if all of them die. The scheduler members are initialized:

```
public class TimeoutSchedulerService
    extends TimeoutSchedulerAbstract {
  private Set<Process> processes = new HashSet<>();
  private ExecutorService monitorService =
    Executors.newCachedThreadPool();

  @Override
  public boolean onSchedule(PackingPlan packing) {...}

  @Override
  public boolean onKill(Scheduler.KillTopologyRequest request)
}
```

The `onSchedule()` method starts all Heron Executor processes and starts the monitoring thread to watch the Heron Executor process states:

```
public boolean onSchedule(PackingPlan packing) {
  processes.add(ShellUtils.runASyncProcess(
      getExecutorCommand(0), workingDirectory, null)); (1)
  packing.getContainers().forEach(cp -> processes.add(ShellUtils
      .runASyncProcess(getExecutorCommand(cp.getId()),
          workingDirectory, null))); (2)
  monitorService.submit(() -> {
```

```
    processes.stream().forEach(p -> {
      try {
        p.waitFor();  ③
      } catch (InterruptedException e) {
        e.printStackTrace();
      }
    });
    monitorService.shutdownNow();  ④
    String topologyName = Context.topologyName(runt);
    Runtime.schedulerStateManagerAdaptor(runt)
        .deleteExecutionState(topologyName);
    Runtime.schedulerStateManagerAdaptor(runt)
        .deleteTopology(topologyName);
    Runtime.schedulerStateManagerAdaptor(runt)
        .deleteSchedulerLocation(topologyName);  ⑤
    Runtime.schedulerShutdown(runt).terminate();  ⑥
  });
  return true;
}
```

① Run Heron Executor in container 0.
② Run Heron Executor in the remaining containers. Since the packing plan does not contain container 0, container 0 is run separately
③ Wait for all Heron Executor processes to die. waitFor() is a blocking call, and the monitorService is blocked here until all Heron Executor processes quit.
④ Shut down the monitorService thread.
⑤ Clean the states in State Manager. Ideally, all states should be removed; however, for simplicity, the example cleans three key states: execution state, topology definition, and scheduler location.
⑥ Shut down the scheduler process.

Since the scheduler process memorizes all the Heron Executor process handlers, killing those processes is easily done by calling the handler's destroy() method:

```
public boolean onKill(Scheduler.KillTopologyRequest request) {
  processes.stream().forEach(p -> p.destroy());
  return true;
}
```

12.2.2.3 Library Mode

Compared to service mode, the onSchedule() of library mode, implemented in *TimeoutSchedulerLibrary.java*, is much simpler. First, it does not maintain any states in the scheduler object because a new scheduler object will be instantiated every time. Second, it does not monitor Heron Executor processes because the scheduler object lifetime is shorter than the Heron Executor processes. It contains only two methods:

```
public class TimeoutSchedulerLibrary
    extends TimeoutSchedulerAbstract {
  @Override
  public boolean onSchedule(PackingPlan packing) {...}

  @Override
  public boolean onKill(Scheduler.KillTopologyRequest request)
}
```

The onSchedule() method simply starts the Heron Executor processes and returns:

```
public boolean onSchedule(PackingPlan packing) {
  ShellUtils.runASyncProcess(
    getExecutorCommand(0), workingDirectory, null);
  packing.getContainers().forEach(
    cp -> ShellUtils.runASyncProcess(
      getExecutorCommand(cp.getId()), workingDirectory, null));
  return true;
}
```

When onKill() is called in a new scheduler object, there has to be a way to find the Heron Executor processes that another scheduler object spawned in the last onSchedule() invocation. Fortunately, each Heron Executor process writes its PID in a file in its container, which can be used to locate the Heron Executor process as long as the container directory is known:

```
public boolean onKill(Scheduler.KillTopologyRequest request) {
  Arrays.stream(workingDirectory.listFiles(
      (d, n) -> n.matches("heron-executor-[0-9]+.pid")))  ①
    .forEach(f -> ShellUtils.runSyncProcess(true, false,
        new String[]{"pkill", "-F", f.getName()},
          null, workingDirectory));  ②
  return true;
}
```

①   Find all Heron Executor process PID files by filtering the filename using a regular expression.

②   Call the **pkill** command to signal TERM to the Heron Executor processes.

12.2.2.4 Service Mode Versus Library Mode

Service mode is the default mode to extend Heron Scheduler. It owns a process to maintain the Heron Executor states during a topology's lifetime. With Heron Executor states or handlers, it can easily achieve update, restart, and kill functions.

Section 12.1.1 discussed the criteria to run the service mode. When the scheduler service runs in container 0, the first criterion in the life cycle is easily satisfied by making their life cycle the same.

The Heron CLI usually has access to the containers directly or through the Heron API server. If there is no access to the containers in service mode, container access cannot be achieved in library mode as well since library mode requires the Heron CLI to manage containers directly.

In some circumstances, the third criterion is hard to achieve, such as Aurora scheduler; thus, the library mode is adopted. For Aurora, if the scheduler process runs inside container 0, the Aurora client has to be available, and the users' privileges have to be delegated to the scheduler process to launch the rest of the containers, which is hard to achieve in some data centers. For resource pools like Aurora and Kubernetes, the library mode is chosen not only due to the practical authentication and authorization issues, but also due to the powerful resource pool controller substituting the scheduler process functions.

Although the service mode is the default, the service mode or library mode for a new scheduler should be decided on a case-by-case basis.

12.3 Summary

This chapter described the Heron Scheduler SPI, especially the `ILauncher` and `IScheduler` interfaces, and how they work together to launch the Heron Executor processes in containers. The two scheduler modes—service mode and library mode—were introduced and compared.

This chapter also showed the procedures to write a customized scheduler: *timeout* scheduler. The example demonstrated not only the critical implementation of SPI interfaces in both service and library modes, but also showed how to organize, configure, and compile the code. The Heron out-of-box schedulers follow the same implementation procedures described in this chapter.

Heron is an evolving project. The community continues to develop new features, some of which are already in version 0.20.3-incubating-rc7. We will take a look at these new features in the next chapter.

Chapter 13
Heron Is Evolving

The Heron project is always evolving. Several new modules have been added, and we can expect more new features and modules in the future, thanks to the active community.

13.1 Dhalion and Health Manager

Dhalion was first proposed by Floratou et al. at the 2017 VLDB conference, based on which the Heron module Health Manager was built. Health Manager is responsible for tuning, stabilizing, and healing a topology.

13.1.1 Dhalion

Stream processing systems have to address some crucial challenges facing their operators [1]:

- The manual, time-consuming, and error-prone tasks of tuning various configurations to achieve service-level objectives (SLOs).
- The maintenance of SLOs with unexpected traffic variation and hardware/software performance degradation.

Motivated by these challenges, Dhalion is a proposed system built on the core philosophy that stream processing systems must self-regulate. Three important capabilities are defined to make a system self-regulating [1]:

Self-tuning
Since there is no principled way to fully determine the ideal configuration, users typically try several configurations and pick the one that best matches their SLOs.

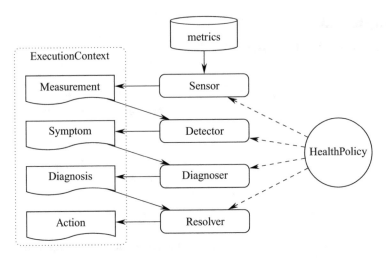

Fig. 13.1 Dhalion architecture

A self-regulating streaming system should take the specification of a streaming application as well as a policy defining the objective and automatically tune configuration parameters to achieve the stated objective.

Self-stabilizing

Since the load variations are largely unpredictable, operators are forced to over-provision resources for these applications to avoid SLO violations. A self-regulating streaming system must react to external shocks by appropriately reconfiguring itself to guarantee stability (and SLO adherence) at all times.

Self-healing

System performance can be affected not only by hardware or software failures but also by hardware and software delivering degraded quality of service, such as a slow disk and memory constraints, leading to swap thrashing. A self-regulating streaming system must identify such service degradations, diagnose the internal faults that are at their root, and perform the necessary actions to recover from them.

Dhalion has a modular and extensible architecture, as shown in Fig. 13.1. Dhalion periodically invokes a policy that has four phases in the following order [1]:

1. Metric collection phase: Dhalion observes the system state by collecting various metrics from the underlying streaming system. This phase generates a collection of Measurements. A Measurement is a data point of five-element tuples <timestamp, value, component, instance, type>.
2. Symptom detection phase: Based on the Measurements collected, Dhalion tries to distinguish Symptoms linked to the streaming application health issues.

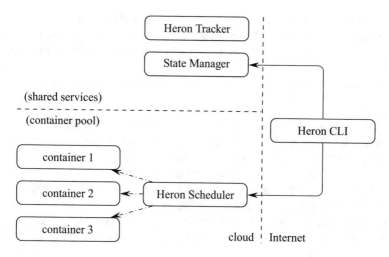

Fig. 13.2 Deployment: library mode

3. Diagnosis generation phase: After collecting various `Symptoms`, Dhalion attempts to find one or more `Diagnoses` that explain them. Dhalion generates all the possible `Diagnoses` that can explain the collected `Symptoms`.
4. Resolution phase: Once `Diagnoses` have been found, the system evaluates them and tries the possible actions to resolve the problem.

The output of policy phases is cached in `ExecutionContext` as "fact" tables ready for queries from a later phase in the same and subsequent policy invocations.

13.1.2 Health Manager

Heron Health Manager is a process inside container 0. It is, following the Dhalion guideline, made of policies of four phases. The metric collection phase often fetches metrics from the MetricsCache Manager process in the same container.

Thanks to the Dhalion pluggable architecture, policies can be easily added to Health Manager. The current policies include `DynamicResourceAllocation Policy` and `AutoRestartBackpressureContainerPolicy`. The idea behind these policies is implementing Site Reliability Engineer (SRE) operations in a policy.

13.2 Deploy Mode (API Server)

In the previous chapters, we saw how to use the Heron CLI to submit topology, which is the library deployment mode. A new deployment mode, service mode, was introduced to work with cloud-native environments such as Kubernetes scheduler. Note that the scheduler mode in Sect. 12.2.2.4 is different from deploy mode. Scheduler mode talks about the scheduler process, while deploy mode describes the Heron CLI.

There were discussions on the *heron-dev* mailing list.[1] Here are some notes:

Library Mode

With Heron, the current mode of deployment is called the library mode, as shown in Fig. 13.2. This mode does not require any services running for Heron to deploy, which is a huge advantage. However, it requires several configurations on the client side. Because of this, administering becomes harder, especially maintaining the configurations and distributing them when a configuration is changed. While this is possible for bigger teams with a dedicated dev-ops team, it might be overhead for medium and smaller teams. Furthermore, this mode of deployment does not have an API to submit/kill/activate/deactivate programmatically.

Service Mode

In this mode, an API server will run as a service. This service will be run as yet another job in the scheduler so that it will be restarted during machine and process failures, thereby providing fault tolerance. This API server will maintain the configuration, and the Heron CLI will be augmented to use the REST API to submit/kill/activate/deactivate the topologies in this mode. The advantage of this mode is that it simplifies deployment but requires running a service.

Merging Heron Tracker into the API server (future)

Currently, Heron Tracker written in Python duplicates the State Manager code in Python as well. The API server will support the Heron Tracker API in addition to the Topologies API. Depending on the mode of the deployment, the API server can be deployed in one of the two modes—library mode (which exposes only the tracker API) or service mode (which exposes both the tracker and the API server). Initially, the tracker and the API server will be in separate directories until a great amount of testing is done. Once completed, we can think about cutting over to entirely using the API server, as shown in Fig. 13.3.

---

[1] http://mail-archives.apache.org/mod_mbox/heron-dev/201707.mbox/thread.

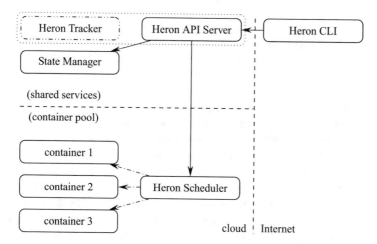

Fig. 13.3 Deployment: service mode (with Tracker)

13.3   Cloud-Native Heron

We saw that Heron could run with multiple schedulers in Chap. 12, among which Kubernetes is a bridge enabling Heron to be cloud-native. In this section, we use a local cloud environment of Minikube with Docker to experiment on how to process a Heron job within the cloud. The following steps demonstrate a simplified full life cycle of cloud-native Heron: compiling, packaging an image, launching a cluster, submitting a job, and recycling resources.

1. Install Docker[2] :

```
$ sudo apt-get install -y \ ①
> apt-transport-https ca-certificates curl \
> gnupg-agent software-properties-common

$ curl -fsSL https://download.docker.com/linux/ubuntu/gpg | \
> sudo apt-key add - ②
$ sudo add-apt-repository \ ③
> "deb [arch=amd64] https://download.docker.com/linux/ubuntu \
> $(lsb_release -cs) \
> stable"

$ sudo apt-get update
$ sudo apt-get install -y \
> docker-ce docker-ce-cli containerd.io ④
$ docker --version ⑤
```

---

[2]https://docs.docker.com/engine/install/ubuntu/#install-using-the-repository.

```
$ sudo groupadd docker
$ sudo usermod -aG docker $USER  (6)
$ newgrp docker  (7)
```

(1)   Get ready for **apt** to install Docker.

(2)   Add Docker's official GPG key.

(3)   Set up the stable repository. The **lsb_release -cs** subcommand returns the name of Ubuntu distribution, such as *focal*.

(4)   Install Docker Engine and containerd.[3]

(5)   Verify that the **docker** command is ready.

(6)   The Docker daemon runs as the *root* user and binds to a Unix socket owned by the user *root* by default. The other users can only access that socket using **sudo**. If you do not want to type **sudo** before **docker**, create a *docker* group and add users to it. When the Docker daemon starts, the Unix socket is accessible by the *docker* group members.

(7)   Activate the changes to groups

2. Compile Heron in a Docker container:

```
$ cd ~/heron && \
> ./docker/scripts/build-artifacts.sh \  (1)
> ubuntu20.04 \  (2)
> 0.0.0 \  (3)
> ~/heron-release  (4)
```

(1)   The script to build Heron artifacts for different platforms.

(2)   The supported target platforms include *darwin*, *debian10*, *ubuntu20.04*, and *centos7*.

(3)   The Heron version. For this experiment, we feed a dummy version.

(4)   The output directory.

3. Install the **kubectl**[4] command:

```
$ VER=$(curl -s https://storage.googleapis.com\
> /kubernetes-release/release/stable.txt)
$ curl -LO https://storage.googleapis.com\  (1)
> /kubernetes-release/release/${VER}/bin/linux/amd64/kubectl

$ chmod +x ./kubectl
$ sudo mv ./kubectl /usr/local/bin/kubectl  (2)

$ kubectl version --client  (3)
```

(1)   Download the latest release.

(2)   Move the binary into your PATH.

(3)   Test the command.

---

[3]https://containerd.io/.

[4]https://kubernetes.io/docs/tasks/tools/install-kubectl/#install-kubectl-binary-with-curl-on-linux.

Install the Minikube[5] tool:

```
$ curl -Lo minikube https://storage.googleapis.com\
> /minikube/releases/latest/minikube-linux-amd64  (1)
$ chmod +x minikube
$ sudo mkdir -p /usr/local/bin/
$ sudo install minikube /usr/local/bin/
$ minikube version

$ minikube start --driver=docker  (2)
$ minikube status  (3)
$ kubectl config current-context
minikube  (4)
```

(1)	Download the stand-alone binary.
(2)	Start a local Kubernetes cluster. The **kubectl** command is now configured to use the minikube context.
(3)	Check the status of the cluster.
(4)	Verify that the **kubectl** context is minikube.

4. Share the Minikube image repository to the **Minikube** command and build an image in the Minikube image repository:

```
$ docker images  (1)
$ eval $(minikube -p minikube docker-env)  (2)
$ docker images  (3)

$ cd ~/heron && \
> ./docker/scripts/build-docker.sh \  (4)
> ubuntu20.04 \  (5)
> 0.0.0 \  (6)
> ~/heron-release  (7)

$ docker images  (8)
```

(1)	The **docker** command points to a default docker-daemon. We used this environment to compile Heron.
(2)	To share the Minikube image repository, point the shell to the Minikube docker-daemon.
(3)	The **docker** command lists images within the Minikube environment.
(4)	The script to build the Heron Docker image for different platforms.
(5)	<platform>: darwin, debian10, ubuntu20.04, and centos7.
(6)	<version_string>: the version of Heron build, e.g., v0.17.5.1-rc. In our experiment, we use the dummy version 0.0.0.
(7)	<artifact-directory>: location of the compiled Heron artifact.
(8)	Verify that the image heron/heron:0.0.0 is ready.

---

[5]https://kubernetes.io/docs/tasks/tools/install-minikube/#installing-minikube.

5. Start a Heron cluster:

```
$ DIR=./deploy/kubernetes/minikube
$ STR='s!heron/heron:latest!heron/heron:0.0.0!g'  ①
$ sed ${STR} ${DIR}/zookeeper.yaml > /tmp/zookeeper.yaml
$ sed ${STR} ${DIR}/tools.yaml > /tmp/tools.yaml
$ sed ${STR} ${DIR}/apiserver.yaml > /tmp/apiserver.yaml

$ kubectl create -f /tmp/zookeeper.yaml ②
$ kubectl create -f ${DIR}/bookkeeper.yaml
$ kubectl create -f /tmp/tools.yaml
$ kubectl create -f /tmp/apiserver.yaml ③
```

①   Set the image version to 0.0.0.

②   ZooKeeper is a dependency of other processes, thus start it first.

③   This YAML file sets the cluster name *kubernetes*.

6. Submit a job:

```
$ kubectl proxy -p 8001 &  ①
$ curl http://localhost:8001/api/v1\
> /namespaces/default/services/heron-apiserver:9000\
> /proxy/api/v1/version ②
{
    "heron.build.git.revision" : "...",
    "heron.build.git.status" : "Clean",
    "heron.build.host" : "vm",
    "heron.build.time" : "Sun Jul 5 01:52:20 UTC 2020",
    "heron.build.timestamp" : "1593914249000",
    "heron.build.user" : "ubuntu",
    "heron.build.version" : "..."
}

$ heron config kubernetes \  ③
> set service_url http://localhost:8001/api/v1\
> /namespaces/default/services/heron-apiserver:9000\
> /proxy
$ heron submit kubernetes \  ④
> ~/.heron/examples/heron-api-examples.jar \
> org.apache.heron.examples.api.AckingTopology \
> acking
```

①   This command starts a proxy to the Kubernetes API server.

②   Test to verify the Heron API server version.

③   Set the *service_url* property of the *kubernetes* cluster.

④   Submit a job named *acking* to the *kubernetes* cluster.

7. Kill the job and stop the cluster:

```
$ heron kill kubernetes acking

$ kubectl delete -f /tmp/zookeeper.yaml
$ kubectl delete -f ${DIR}/bookkeeper.yaml
$ kubectl delete -f /tmp/tools.yaml
$ kubectl delete -f /tmp/apiserver.yaml
```

13.4 Summary

This chapter presented the Health Manager module to maintain the job SLO proactively. The Heron API server is necessary for some application scenarios, such as on Kubernetes. Although Heron has already provided many useful features, it keeps going to provide more and better features.

Reference

1. Floratou, A., Agrawal, A., Graham, B., Rao, S., Ramasamy, K.: Dhalion: Self-regulating stream processing in Heron. Proceedings of the VLDB Endowment **10**(12), 1825–1836 (2017)

Index

A
At-least-once, 16, 81
At-most-once, 16
Apache Flink, 12
Apache Software Foundation (ASF), 11
Apache Storm, 8
Application programming interface (API), ix, 6

B
Batch processing, 3
Bazel, 28
Bolt, 18
BUILD, 39

C
Central communication bus, 23
Clone, 104
Command-line interface (CLI), 15

D
Dhalion, 197
Directed acyclic graph (DAG), 4

E
Effectively-once, 16, 84
Exactly-once, 85
Extensible Component Orchestrator (ECO), 58

F
Filter, 101
FlatMap, 101

Fully distributed communication, 22

G
Grouping, 18

H
Hadoop distributed file system (HDFS), 3
Heron CLI, 20
Heron Instance, 22
Heron Tracker, 20
Heron UI, 20

I
Internet of things (IoT), 3

J
Java archive (JAR), 9
Java virtual machine (JVM), 9
Join, 106

K
Kappa architecture, 6

L
Lambda architecture, 4
Logical plan, 22

M
Map, 101

Printed in the United States
by Baker & Taylor Publisher Services